Space
and
Place
*The Perspective
of
Experience*

Space
and
Place

The Perspective
of
Experience

□ Yi-Fu Tuan □

University of Minnesota Press
Minneapolis
London

Published by the University of Minnesota Press
111 Third Avenue South, Suite 290
Minneapolis, MN 55401-2520

Library of Congress Catalog Number 77-072910

ISBN 0-8166-3877-2

Fifth Printing, 2007

Preface

The life of thought is a continuous story, like life itself: one book grows out of another as in the world of political commitment one action leads to another. I wrote a book called *Topophilia* out of the need to sort and order in some way the wide variety of attitudes and values relating to man's physical environment. While I enjoyed noting the richness and range of human environmental experience, I could not at that time find an overarching theme or concept with which to structure my heterogeneous material; and in the end I often had to resort to convenient and conventional categories (like suburb, town, and city, or the separate treatment of the human senses) rather than to categories that evolved logically out of a ruling theme. The present book is an attempt to achieve a more coherent statement. To do this I narrow my focus to the closely related "space" and "place" components of environment. More importantly, I try to develop my material from a single perspective — namely, that of experience. The complex nature of human experience, which ranges from inchoate feeling to explicit conception, commands the subject matter and themes of this book.

It has often been difficult for me to acknowledge properly my intellectual debts. One reason is that I owe so much to so many. An even greater problem is that I may well fail to acknowledge people to whom I owe the greatest debt. I have cannibalized them! Their ideas have become my own innermost thoughts. My unnamed mentors include students and colleagues at the University of Minnesota. I expect them to be indulgent toward any unconscious borrowing of their insights, for all teachers know it to be the sincerest form of compliment.

I do have specific debts, and it gives me pleasure to acknowledge them. I am deeply grateful to J. B. Jackson and P. W. Porter for their encouragement of my fumbling efforts; to Su-chang Wang, Sandra Haas, and Patricia Burwell for the diagrams which achieve a formal elegance that in the case of the text remains only an aspiration; and to Dorian Kottler of the University Press for the meticulous job of copyediting. I also want to thank the following institutions, which provided me with the resources to work on *Space and Place* with little interruption in the last two years: the University of Minnesota for granting me a sabbatical leave followed by a year's leave of absence; the University of Hawaii, where I first explored the themes of this book with a small group of sympathetic graduate students; the Australian-American Educational Foundation (Fulbright-Hays program) for an award to visit Australia; the Department of Human Geography at the Australian National University for providing a congenial and stimulating environment in which to think and write; and the University of California at Davis for a year of sunshine and warmth, human and climatic.

Yi-Fu Tuan

Chinese New Year, 1977

Contents

Illustrations

Space
and
Place
*The Perspective
of
Experience*

1

Introduction

"**S**pace" and "place" are familiar words denoting common experiences. We live in space. There is no space for another building on the lot. The Great Plains look spacious. Place is security, space is freedom: we are attached to the one and long for the other. There is no place like home. What is home? It is the old homestead, the old neighborhood, hometown, or motherland. Geographers study places. Planners would like to evoke "a sense of place." These are unexceptional ways of speaking. Space and place are basic components of the lived world; we take them for granted. When we think about them, however, they may assume unexpected meanings and raise questions we have not thought to ask.

What is space? Let an episode in the life of the theologian Paul Tillich focus the question so that it bears on the meaning of space in experience. Tillich was born and brought up in a small town in eastern Germany before the turn of the century. The town was medieval in character. Surrounded by a wall and administered from a medieval town hall, it gave the impression of a small, protected, and self-contained world. To an imaginative child it felt narrow and restrictive. Every year, however young Tillich was able to escape with his family to the Baltic Sea. The flight to the limitless horizon and unrestricted space

3

of the seashore was a great event. Much later Tillich chose a place on the Atlantic Ocean for his days of retirement, a decision that undoubtedly owed much to those early experiences. As a boy Tillich was also able to escape from the narrowness of small-town life by making trips to Berlin. Visits to the big city curiously reminded him of the sea. Berlin, too, gave Tillich a feeling of openness, infinity, unrestricted space.[1] Experiences of this kind make us ponder anew the meaning of a word like "space" or "spaciousness" that we think we know well.

What is a place? What gives a place its identity, its aura? These questions occurred to the physicists Niels Bohr and Werner Heisenberg when they visited Kronberg Castle in Denmark. Bohr said to Heisenberg:

Isn't it strange how this castle changes as soon as one imagines that Hamlet lived here? As scientists we believe that a castle consists only of stones, and admire the way the architect put them together. The stones, the green roof with its patina, the wood carvings in the church, constitute the whole castle. None of this should be changed by the fact that Hamlet lived here, and yet it is changed completely. Suddenly the walls and the ramparts speak a quite different language. The courtyard becomes an entire world, a dark corner reminds us of the darkness in the human soul, we hear Hamlet's "To be or not to be." Yet all we really know about Hamlet is that his name appears in a thirteenth-century chronicle. No one can prove that he really lived, let alone that he lived here. But everyone knows the questions Shakespeare had him ask, the human depth he was made to reveal, and so he, too, had to be found a place on earth, here in Kronberg. And once we know that, Kronberg becomes quite a different castle for us.[2]

Recent ethological studies show that nonhuman animals also have a sense of territory and of place. Spaces are marked off and defended against intruders. Places are centers of felt value where biological needs, such as those for food, water, rest, and procreation, are satisfied. Humans share with other animals certain behavioral patterns, but as the reflections of Tillich and Bohr indicate, people also respond to space and place in complicated ways that are inconceivable in the animal world. How can the Baltic Sea and Berlin both evoke a sense of openness and infinitude? How can a mere legend haunt Kronberg Castle and impart a mood that infiltrates the minds of two famous scientists? If our concern with the nature and quality of the human environment is serious, these are surely basic questions. Yet they have seldom been raised. Instead we study animals such as rats and wolves and say that human behavior and values are much like theirs. Or we measure and map space and place, and acquire spatial laws and resource inventories for our efforts. These are important approaches, but they need to be complemented by experiential data that we can collect and interpret in measured confidence because we are human ourselves. We have privileged access to states of mind, thoughts and feelings. We have an insider's view of human facts, a claim we cannot make with regard to other kinds of facts.

People sometimes behave like cornered and wary animals. On occasion they may also act like cool scientists dedicated to the task of formulating laws and mapping resources. Neither posture holds sway for long. People are complex beings. The human endowment includes sensory organs similar to those of other primates, but it is capped by an exceptionally refined capacity for symbolization. How the human person, who is animal, fantasist, and computer combined, experiences and understands the world is the central theme of this book.

Given the human endowment, in what ways do people attach meaning to and organize space and place? When this question is asked, the social scientist is tempted to rush to culture as an explanatory factor. Culture is uniquely developed in human beings. It strongly influences human behavior and values. The Eskimos' sense of space and place is very different from that of Americans. This approach is valid, but it overlooks the problem of shared traits that transcend cultural particularities and may therefore reflect the general human condition. When note is taken of "universals," the behavioral scientist is likely to turn to the analogue of primate behavior. In this book our animal heritage is assumed. The importance of culture is taken for granted; culture is inescapable, and it is explored in every chapter. But the purpose of the essay is not to produce a handbook of how cultures affect human attitudes to space and place. The essay is, rather, a prologue to culture in its countless variety; it focuses on general questions of human dispositions,

Introduction

capacities, and needs, and on how culture emphasizes or distorts them. Three themes weave through the essay. They are:

(1) The biological facts. Human infants have only very crude notions of space and place. In time they acquire sophistication. What are the stages of learning? The human body lies prone, or it is upright. Upright it has top and bottom, front and back, right and left. How are these bodily postures, divisions, and values extrapolated onto circumambient space?

(2) The relations of space and place. In experience, the meaning of space often merges with that of place. "Space" is more abstract than "place." What begins as undifferentiated space becomes place as we get to know it better and endow it with value. Architects talk about the spatial qualities of place; they can equally well speak of the locational (place) qualities of space. The ideas "space" and "place" require each other for definition. From the security and stability of place we are aware of the openness, freedom, and threat of space, and vice versa. Furthermore, if we think of space as that which allows movement, then place is pause; each pause in movement makes it possible for location to be transformed into place.

(3) The range of experience or knowledge. Experience can be direct and intimate, or it can be indirect and conceptual, mediated by symbols. We know our home intimately; we can only know *about* our country if it is very large. A longtime resident of Minneapolis knows the city, a cabdriver learns to find his way in it, a geographer studies Minneapolis and knows the city conceptually. These are three kinds of experiencing. One person may know a place intimately as well as conceptually. He can articulate ideas but he has difficulty expressing what he knows through his senses of touch, taste, smell, hearing, and even vision.

People tend to suppress that which they cannot express. If an experience resists ready communication, a common response among activists ("doers") is to deem it private—even idiosyncratic—and hence unimportant. In the large literature on environmental quality, relatively few works attempt to understand how people feel about space and place, to take into account the different modes of experience (sensorimotor, tactile, visual, conceptual), and to interpret space and place as images of complex—often ambivalent—feelings. Professional planners, with their urgent need to act, move too quickly to models and inventories. The layman accepts too readily from charismatic planners and propagandists the environmental slogans he may have picked up through the media; the rich experiential data on which these abstractions depend are easily forgotten. Yet it is possible to articulate subtle human experiences. Artists have tried—often with success. In works of literature as well as in humanistic psychology, philosophy, anthropology and geography, intricate worlds of human experience are recorded.

This book draws attention to questions that humanists have posed with regard to space and place.[3] It attempts to systematize humanistic insights, to display them in conceptual frames (here organized as chapters) so that their importance is evident to us not only as thoughtful people curious to know more about our own nature—our potential for experiencing—but also as tenants of the earth practically concerned with the design of a more human habitat. The approach is descriptive, aiming more often to suggest than to conclude. In an area of study where so much is tentative, perhaps each statement should end with a question mark or be accompanied by qualifying clauses. The reader is asked to supply them. An exploratory work such as this should have the virtue of clarity even if this calls for the sacrifice of scholarly detail and qualification.

A key term in the book is "experience." What is the nature of experience and of the experiential perspective?

2

Experiential
Perspective

Experience is a cover-all term for the various modes through which a person knows and constructs a reality. These modes range from the more direct and passive senses of smell, taste, and touch, to active visual perception and the indirect mode of symbolization.[1]

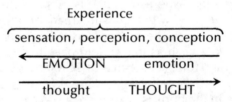

Emotion tints all human experience, including the high flights of thought. Mathematicians, for example, claim that the design of their theorems is guided by aesthetic criteria— notions of elegance and simplicity that answer a human need. Thought tints all human experience, including the basic sensations of heat and cold, pleasure and pain. Sensation is quickly qualified by thought as one of a special kind. Heat is suffocating or prickly; pain is sharp or dull, an irritating tease or a brutal force.

Experience is directed to the external world. Seeing and thinking clearly reach out beyond the self. Feeling is more ambiguous. As Paul Ricoeur put it, "Feeling is . . . without doubt intentional: it is a feeling of 'something'—the lovable, the hateful, [for instance]. But it is a very strange intentionality which on the one hand designates qualities felt *on* things, *on* persons, *on* the world, and on the other hand manifests and reveals the way in which the self is inwardly affected." In feeling "an intention and an affection coincide in the same experience."[2]

Experience has a connotation of passivity; the word suggests what a person has undergone or suffered. An experienced man or woman is one to whom much has happened. Yet we do not speak of the plant's experiences, and even of the lower animals the word "experience" seems inappropriate. The young pup, however, is contrasted with the experienced mastiff; and human beings are mature or immature depending on whether they have benefited from events. Experience thus implies the ability to learn from what one has undergone.[3] To experience is to learn; it means acting on the given and creating out of the given. The given cannot be known in itself. What can be known is a reality that is a construct of experience, a creation of feeling and thought. As Susanne Langer put it: "The world of physics is essentially the real world construed by mathematical abstractions, and the world of sense is the real world construed by the abstractions which the sense organs immediately furnish."[4]

Experience is the overcoming of perils. The word "experience" shares a common root (*per*) with "experiment," "expert," and "perilous."[5] To experience in the active sense requires that one venture forth into the unfamiliar and experiment with the elusive and the uncertain. To become an expert one must dare to confront the perils of the new. Why should one so dare? A human individual is driven. He is passionate, and passion is a token of mental force. The emotional repertoire of a clam is very restricted compared with that of a puppy; and the affective life of the chimpanzee seems almost as varied and intense as that of a human being. A human infant is distinguished from other mammalian young both by his helplessness and by his fearsome tantrums. The infant's emotional range, from smile to tantrum, hints at his potential intellectual reach.

Experience is compounded of feeling and thought. Human feeling is not a succession of discrete sensations; rather memory and anticipation are able to wield sensory impacts into a shifting stream of experience so that we may speak of a life of feeling as we do of a life of thought. It is a common tendency to regard feeling and thought as opposed, the one registering subjective states, the other reporting on objective reality. In fact, they lie near the two ends of an experiential continuum, and both are ways of knowing.

To see and to think are closely related processes. In English, "I see" means "I understand." Seeing, it has long been recognized, is not the simple recording of light stimuli; it is a selective and creative process in which environmental stimuli are organized into flowing structures that provide signs meaningful to the purposive organism. Are the senses of smell and touch informed by mentality? We tend to slight the cognitive power of these senses. Yet the French verb "savoir" (to know) is closely related to the English "savour." Taste, smell, and touch are capable of exquisite refinement. They discriminate among the wealth of sensations and articulate gustatory, olfactory, and textural worlds.

The structuring of worlds calls for intelligence. Like the intellectual acts of seeing and hearing, the senses of smell and touch can be improved with practice so as to discern significant worlds. Human adults can develop extraordinary sensitivity to a wide range of flower fragrances.[6] Although the human nose is far less acute than the canine nose in detecting certain odors of low intensity, people may be responsive to a broader range of odors than dogs are. Dogs and young children do not appreciate flower fragrances in the way human adults do. Young children's favorite odors are those of fruits rather than flowers.[7] Fruits are good to eat, so preference for them is understandable. But what is the survival value of sensitivity to the chemical oils wafted by flowers? No clear biological purpose is served by this sensitivity. It would seem that our nose, no less than our eyes, seeks to enlarge and comprehend the world. Some odors do have potent biological meaning. Body scents, for example, may stimulate sexual activity. Why, on the other hand, do many human adults find the smell of decay repulsive? Mammals with noses far keener than the human tolerate and even appreciate carrion odors that would disgust men. Young children also appear to be indifferent to fetid smells. Langer suggests that the odors of decay are *memento mori* to grown people but carry no such meaning to animals and small children.[8] Touch articulates another kind of complex world. The human hand is peerless in its strength, agility, and sensitivity. Primates, including man, use their hands to know and comfort members of their own species, but man also uses hands to explore the physical environment, carefully differentiating it by the feel of bark and stone.[9] Human adults dislike having sticky matter on their skin, perhaps because it destroys the skin's power for discernment. Such a substance, like dirty spectacles, dulls a faculty of exploration.

The modern architectural environment may cater to the eye, but it often lacks the pungent personality that varied and pleasant odors can give. Odors lend character to objects and places, making them distinctive, easier to identify and remember. Odors are important to human beings. We have even spoken of an olfactory world, but can fragrances and scents constitute a world? "World" suggests spatial structure; an olfactory world would be one where odors are spatially disposed, not simply one in which they appear in random succession or as inchoate mixtures. Can senses other than sight and touch provide a spatially organized world? It is possible to argue that taste, odor, and even hearing cannot in themselves give us a sense of space.[10] The question is largely academic, for most people function with the five senses, and these constantly reinforce each other to provide the intricately ordered and emotion-charged world in which we live. Taste, for example, almost invariably involves touch and smell: the tongue rolls around the hard candy, exploring its shape as the olfactory sense registers the caramel flavor. If we can hear and smell something we can often also see it.

What sensory organs and experiences enable human beings to have their strong feeling for space and for spatial qualities? Answer: kinesthesia, sight, and touch.[11] Movements such as the simple ability to kick one's legs and stretch one's arms are basic to the awareness of space. Space is experienced directly as having room in which to move. Moreover, by shifting from one place to another, a person acquires a sense of direction. Forward, backward, and sideways are experientially differentiated, that is, known subconsciously in the act of motion. Space assumes a rough coordinate frame centered on the mobile and purposive self. Human eyes, which have bifocal overlap and stereoscopic capacity, provide people with a vivid space in three dimensions. Experience, however, is necessary. It takes time and practice for the infant or the person born blind but with sight recently restored to perceive the world as made up of stable three-dimensional objects arranged in space rather than as shifting patterns and colors. Touching and manipulating things with the hand yields a world of objects— objects that retain their constancy of shape and size. Reaching for things and playing with them disclose their separateness and relative spacing. Purposive movement and perception, both visual and haptic, give human beings their familiar world of disparate objects in space. Place is a special kind of object. It is a concretion of value, though not a valued thing that can be handled or carried about easily; it is an object in which one can dwell. Space, we have noted, is given by the ability to move. Movements are often directed toward, or repulsed by, objects and places. Hence space can be variously experienced as the relative location of objects or places, as the distances and expanses that separate or link places, and—more abstractly—as the area defined by a network of places (Fig. 1).

Taste, smell, skin sensitivity, and hearing cannot individually (nor perhaps even together) make us aware of a spacious external world inhabited by objects. In combination with the "spatializing" faculties of sight and touch, however, these essentially nondistancing senses greatly enrich our apprehension of the world's spatial and geometrical character. Taste labels some flavors "sharp," others "flat." The meaning of these

Experiential Perspective

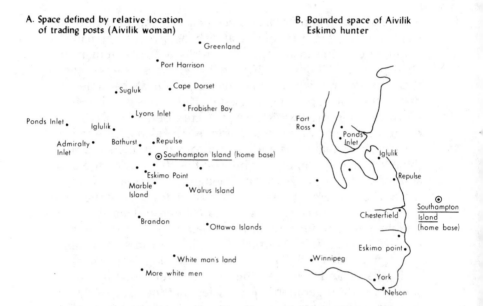

A. Space defined by relative location
of trading posts (Aivilik woman)

B. Bounded space of Aivilik
Eskimo hunter

Figure 1. Space as relative location and bounded space. The Eskimo (Aivilik) woman's space is essentially defined by the location and distance of significant points, mostly trading posts (A), as perceived from the home base on Southampton Island, whereas the idea of boundary (the coastline) is important to the male Eskimo's sense of space (B). Edmund Carpenter, Frederick Varley, and Robert Flaherty, *Eskimo* (Toronto: University of Toronto Press, 1959), page 6. Reprinted with permission from the University of Toronto Press.

geometrical terms is enhanced by their metaphorical use in the realm of taste. Odor is capable of suggesting mass and volume. Some odors, like musk or tuberosa, are "heavy," whereas others are "delicate," "thin," or "light." Carnivores depend on their acute sense of smell to track down prey, and it may be that their nose is capable of articulating a spatially structured world—at least one that is differentiated by direction and distance. The human nose is a much atrophied organ. We depend on the eye to locate sources of danger and appeal, but with the support of a prior visual world the human nose too can discern direction and estimate relative distance through the strength of an odor.

Experiential Perspective

A person who handles an object feels not only its texture but its geometric properties of size and shape. Apart from manipulation, does skin sensitivity itself contribute to the human spatial experience? It does, though in limited ways. The skin registers sensations. It reports on its own state and at the same time that of the object pressing against it. The skin is not, however, a distance sensor. In this respect tactile perception is at the opposite extreme of visual perception. The skin is able to convey certain spatial ideas and can do so without the support of the other senses, depending only on the structure of the body and the ability to move. Relative length, for example, is registered when different parts of the body are touched at the same time. The skin can convey a sense of volume and mass. No one doubts that "entrance into a warm bath gives our skin a more massive feeling than the prick of a pin."[12] The skin, when it comes in contact with flattish objects, can judge approximately their shape and size. At the micro level, roughness and smoothness are geometric properties that the skin easily recognizes. Objects are also hard or soft. Tactile perception differentiates these characteristics on spatio-geometric evidence. Thus a hard object retains its shape under pressure whereas a soft object does not.[13]

Is a sense of distance and of space created out of the ability to hear? The world of sound would appear to be spatially structured, though not with the sharpness of the visual world. It is possible that the blind man who can hear but has no hands and can barely move lacks all sense of space; perhaps to such a person all sounds are bodily sensations and not cues to the character of an environment. Few people are so severely handicapped. Given sight and the power to move and handle things, sounds greatly enrich the human feeling for space. Human ears are not flexible, so they are less equipped to discern direction than, say, the ears of a wolf. But by turning the head a person can roughly tell the direction of sounds. People are subconsciously aware of the sources of noise, and from such awareness they construe auditory space.

Sounds, though vaguely located, can convey a strong sense of size (volume) and of distance. For example, in an empty cathedral the sound of footsteps tapping sharply on the stone floor creates an impression of cavernous vastness. As for the power of sound to evoke distance, Albert Camus wrote: "In Algeria dogs can be heard barking at night over distances ten times greater than in Europe. The noise thus takes on a nostalgia unknown in our cramped countries."[14] Blind people develop an acute sensitivity to sounds; they are able to use them and their reverberations to evaluate an environment's spatial character. People who can see are less sensitive to auditory cues because they are not so dependent on them. All human beings learn, however, to relate sound to distance in the act of speaking. We alter our tone of voice from soft to loud, from intimate to public, in accordance with the perceived physical and social distances between ourselves and others. The volume and phrasing of our voice as well as what we try to say are constant reminders of proximity and distance.

Sound itself can evoke spatial impressions. The reverberations of thunder are voluminous; the squeaking of chalk on slate is "pinched" and thin. Low musical tones are voluminous whereas those of high pitch seem thin and penetrating. Musicologists speak of "musical space." Spatial illusions are created in music quite apart from the phenomenon of volume and the fact that movement logically involves space.[15] Music is often said to have form. Musical form may generate a reassuring sense of orientation. To the musicologist Roberto Gerhard, "form in music means knowing at every moment exactly where one is. Consciousness of form is really a sense of orientation."[16]

The various sensory spaces bear little likeness to each other. Visual space, with its vividness and size, differs strikingly from diffuse auditory and tactile-sensorimotor spaces. A blind man whose knowledge of space derives from auditory and tactile cues cannot, for some time, appreciate the visual world when he gains sight. The vaulted interior of a cathedral and the sensation of slipping into a warm bath both signify volume or spaciousness, although the experiences are hardly comparable. Likewise the meaning of distance is as varied as its experiential modes: we acquire the feel of distance by the effort of

Experiential Perspective

moving from one place to another, by the need to project our voice, by hearing the dogs bark at night, and by recognizing the environmental cues for visual perspective.

The organization of human space is uniquely dependent on sight. Other senses expand and enrich visual space. Thus sound enlarges one's spatial awareness to include areas behind the head that cannot be seen. More important, sound dramatizes spatial experience. Soundless space feels calm and lifeless despite the visible flow of activity in it, as in watching events through binoculars or on the television screen with the sound turned off, or being in a city muffled in a fresh blanket of snow.[17]

Human spaces reflect the quality of the human senses and mentality. The mind frequently extrapolates beyond sensory evidence. Consider the notion of vastness. The vastness of an ocean is not directly perceived. "We think the ocean as a whole," says William James, "by multiplying mentally the impression we get at any moment when at sea."[18] A continent separates New York from San Francisco. A distance of this order is apprehended through numerical or verbal symbols computed, for example, in days' journeys. "But the symbol will often give us the emotional effect of the perception. Such expressions as the abysmal vault of heaven, the endless expanse of ocean, etc., summarize many computations of the imagination, and give the sense of enormous horizon." Someone with the mathematical imagination of Blaise Pascal will look at the sky and be appalled by its infinite expanse. Blind men are able to know the meaning of a distant horizon. They can extrapolate from their experience of auditory space and of freedom in movement to envisage in their minds' eyes panoramic views and boundless space. A blind man told William James that "he thought few seeing people could enjoy the view from a mountain top more than he."[19]

The mind discerns geometric designs and principles of spatial organization in the environment. For example, Dakota Indians find evidence of circular forms in nature nearly everywhere, from the shape of birds' nests to the course of the stars. In contrast, the Pueblo Indians of the American Southwest tend to see spaces of rectangular geometry. These are examples of the construed space, which depends on the power of the mind to extrapolate far beyond the sense data. Such spaces lie at the conceptual end of the experiential continuum. Three principal types, with large areas of overlap, exist—the mythical, the pragmatic, and the abstract or theoretical. Mythical space is a conceptual schema, but it is also pragmatic space in the sense that within the schema a large number of practical activities, such as the planting and harvesting of crops, are ordered. A difference between mythical and pragmatic space is that the latter is defined by a more limited set of economic activities. The recognition of pragmatic space, such as belts of poor and rich soil, is of course an intellectual achievement. When an ingenious person tries to describe the soil pattern cartographically, by means of symbols, a further move toward the conceptual mode occurs. In the Western world systems of geometry—that is, highly abstract spaces—have been created out of primal spatial experiences. Thus sensorimotor and tactile experiences would seem to lie at the root of Euclid's theorems concerning shape congruence and the parallelism of distant lines; and visual perception is the basis for projective geometry.

Human beings not only discern geometric patterns in nature and create abstract spaces in the mind, they also try to embody their feelings, images, and thoughts in tangible material. The result is sculptural and architectural space, and on a large scale, the planned city. Progress here is from inchoate feelings for space and fleeting discernments of it in nature to their public and material reification.

Place is a type of object. Places and objects define space, giving it a geometric personality. Neither the newborn infant nor the man who gains sight after a lifetime of blindness can immediately recognize a geometric shape such as a triangle. The triangle is at first "space," a blurred image. Recognizing the triangle requires the prior identification of corners—that is, places. A neighborhood is at first a confusion of images to the new resident; it is blurred space "out there." Learning to know the neighborhood requires the identification of significant localities, such as street corners and architectural landmarks, within the neighborhood space. Objects and places are centers of value. They attract or repel in finely shaded degrees. To attend to them even momentarily is to acknowledge their reality and value. The infant's world lacks permanent objects, being dominated by fleeting impressions. How do impressions, given to us through the senses, acquire the stability of objects and places?

Intelligence is manifest in different types of achievement. One is the ability to recognize and feel deeply about the particular. In distinction to the schematic worlds in which animals live, the schematic worlds of human beings are also richly populated with particular and enduring things. The particular things we value may be given names: a tea set is Wedgewood and a chair is Chippendale. People have proper names. They are particular things and they may well be the first permanent objects in the infant's world of unstable impressions. An object such as a valued crystal glass is recognized by its unique shape, decorative design, and ring when lightly tapped. A city such as San Francisco is recognized by its unique setting, topography, skyline, odors, and street noises.[20] An object or place achieves concrete reality when our experience of it is total, that is, through all the senses as well as with the active and reflective mind. Long residence enables us to know a place intimately, yet its image may lack sharpness unless we can also see it from the outside and reflect upon our experience. Another place may lack the weight of reality because we know it only from the outside—through the eyes as tourists, and from reading about it in a guidebook. It is a characteristic of the symbol-making human species that its members can become passionately attached to places of enormous size, such as a nation-state, of which they can have only limited direct experience.

3

Space, Place, and the Child

Feelings and ideas concerning space and place are extremely complex in the adult human being. They grow out of life's unique and shared experiences. Every person starts, however, as an infant. From the infant's tiny and confused world appears in time the adult's world view, subliminally also confused, but sustained by structures of experience and conceptual knowledge. Although children come under cultural influences as soon as they are born, the biological imperatives of growth nonetheless impose rising curves of learning and understanding that are alike and hence may be said to transcend the specific emphases of culture.

How does a young child perceive and understand his environment? Fairly dependable answers are available. The child's biological equipment, for instance, gives clues as to the limits of his powers. Moreover we can observe how the child behaves in controlled and real-life situations. We may also wonder, what is the feeling tone of the child's world? what is the nature of his attachments to people and to places? Such questions are more difficult to answer. An introspective return to our own childhood is often disappointing, for the bright and dark landscapes of our early years tend to fade while only a few landmarks such as birthdays and the first day at school remain.

Space, Place, and
the Child

This inability, for most people, to recapture the mood of their
own childhood world suggests how far the adult's schemata,
geared primarily to life's practical demands, differ from those
of the child.[1] Yet the child is father to the man, and the adult's
perceptual categories are from time to time infused with emotions that surge out of early experiences. These highly charged
moments from the past are sometimes captured by poets. Like
candid snapshots out of the family album their words recall for
us a lost innocence and a lost dread, an immediacy of experience that had not yet suffered (or benefited) from the distanc-
ing of reflective thought.

Biology conditions our perceptual world. At birth an infant's
cerebral cortex has only about 10 to 20 percent of the normal
complement of nerve cells in a mature brain; moreover many
of the nerve cells present are not connected with each other.[2]
The infant has no world. He cannot distinguish between self and
an external environment. He feels, but his sensations are not
localized in space. The pain is simply there, and he responds to
it with crying; he does not seem to locate it in some specific
part of his body. For only a brief time, as infants, human beings
have known how it feels to live in a nondualistic world.

During the first few weeks of life the infant's eyes cannot
focus properly. Toward the end of the first month the infant is
able to fixate an object in the direct line of his vision, and by
the end of the second month binocular fixation with con-
vergence begins to appear.[3] However, even in his fourth
month the infant shows little interest in exploring the world
visually beyond the range of three feet.[4] An infant is immobile
and can make only small movements with his head and limbs.
Moving the body along a more or less straight line is essential
to the experiential construction of space into the basic coordinates of ahead, behind, and sideways. Most mammals, soon
after birth, gain a sense of orientation by taking a few steps
after their mother. The slow-maturing human child must acquire this skill more gradually.

What events and activities can provide the infant with the
feel of space? An infant in the Western world spends much of
his time prone. He is occasionally picked up to be burped,

played with, and consoled. Out of these events may come the felt distinction between horizontal and vertical. At the level of activity an infant knows space because he can move his limbs: kicking aside the blanket that encumbers him is a taste of freedom that, in the adult, is associated with the idea of having space. An infant explores the environment with his mouth.[5] The mouth adjusts to the contour of the mother's breast. Sucking is a most rewarding activity, for it requires participation by the different senses of touch, smell, and taste. In addition, sucking feeds the baby, giving him a sense of contentment. The stomach distends and contracts as food is taken in and digested. This physiological function, unlike breathing, is consciously identified with alternating states of distress and bliss. "Empty" and "full" are visceral experiences of lasting importance to the human being. The infant knows them and responds with crying or smiling. To the adult, such commonplace experiences take on an extra metaphorical meaning, as the expressions "my cup runneth over," "I have an empty feeling," and "a full life" suggest. The infant uses his hands to explore the tactile and geometrical characteristics of his environment. While the mouth tackles the nipple and acquires the feel of buccal space, the hands move busily over the breast. Long before the infant's eyes can focus on a small object and discern its shape his hands will have grasped it and known its physical properties through touch.

The visual world of an infant is peculiarly difficult to describe because we are tempted to assign to it the well-known categories of an adult's visual world. How the senses of smell, taste, and touch structure the environment escapes us most of the time; even educated adults lack a varied vocabulary to present olfactory and tactile worlds. But we have no problem with the visual. Pictures and diagrams, as well as words, come to our aid. The world seen through an adult's or older child's eyes is large and vivid; objects in it are clearly ordered in space. Such is not the case for the infant. His visual space lacks structure and permanence. Objects in it are impressions; hence they tend to exist for the infant only so long as they stay in his visual field.[6] The shapes and sizes of objects lack the constancy that older children take for granted. Piaget notes that an infant may fail to recognize a milk bottle when the wrong end is offered to him; he learns to turn it around when he is about eight months old.[7] To an experienced older child an object looks smaller at a distance, and the diminishment in size of a retreating object is unreflectively transcribed to mean increasing distance. To the infant, however, an object that looks small because it is at a distance may be taken as a different object. The infant does possess an innate capacity to recognize the rough three-dimensional quality of things, their constancy of size and shape, and the distinction between far and near, but the recognition operates within a highly circumscribed field compared with that of a mobile toddler.[8]

The ability to see is strongly supported by nonvisual experiences. Even to an older child the moon overhead is easily considered a different object from the moon on the horizon. That the moon moves around the earth is an abstraction alien to the child's experience: the moon is seen only at specific moments, separated by an interval of time that to the child feels almost eternal. The picture of a road leading to a distant cottage seems easy to interpret; yet the road makes full sense only to someone who has walked on it. An immobile infant can have no sense of distance as the expenditure of energy to overcome spatial barrier. A child quickly learns, however, to read spatial and environmental cues even when they are presented to him in the transcribed form of a picture. A bookish youngster three or four years old can already look at the picture of a footpath disappearing into the woods and see himself as the hero of an impending venture.

The first environment an infant explores is his parent. The first permanent and independent object he recognizes is perhaps another person. While things appear and continue to exist only insofar as he attends to them, the independent reality of an adult, able to dispense or withhold favor, soon intrudes on the child's quickening consciousness.[9] Adults are necessary not only for the child's biological survival, but also for developing his sense of an objective world. An infant a few weeks old has already learned to heed the human presence.

Space, Place, and
the Child

He begins to acquire a sense of distance and direction through the need to judge where a grownup may be. Toward the end of the first month of life an infant is likely to follow with his eyes only one distant percept—the grownup's face. A hungry and crying baby calms down and opens his mouth or makes sucking movements when he sees an adult approaching.

An eight-month-old child is aware of noises, particularly animal and human noises, in the next room. He attends to them; his sphere of interest expands beyond what is visible and of pressing concern. However, his behavioral space remains small. He seems easily discouraged by perceived barriers. According to Spitz, up to about eight months a child's spatial horizon is limited by the bars of his crib, or cot. "Within his cot he grabs toys with ease. If the same toy is offered to him outside the bars of his cot, he reaches for it, but his hands stop at the bars; he does not continue his movements beyond; he could easily do it, for the bars are sufficiently widely spaced. It is as if space ended within his cot. Two or three weeks after the eighth month, however, he suddenly sees the light and becomes able to continue his movement beyond the bars and to grasp his toy."[10]

A crawling baby can explore space. Movement beyond the immediate vicinity of the mother or outside the crib entails risks with which the baby is not prepared to cope. Instincts for survival are not well developed. One that appears between the sixth and the eighth month is fear of the stranger. Prior to this stage the infant makes no distinction between familiar and unfamiliar faces; thereafter he turns his head or cries when a stranger approaches.[11] The inanimate environment provides few unambiguous signals of danger to the intrepid infant explorer. Anything that can be grasped is grasped or put into his mouth for more intimate acquaintance; fear of fire and water has to be learned. To the crawling child horizontal space looks safe. He is aware of one kind of danger in the physical environment: the cliff. Experiments have shown that a baby will not crawl onto a glass plate that extends over a pit with vertical sides despite encouragement from the mother. His eyes respond to cues for sudden changes of slope.[12]

Space, Place, and
the Child

The young child, as soon as he learns to walk, will want to follow his mother and explore the environment within her ambience. The more hostile the environment, the closer the attachment to the protective adult. Bushman babies of southwest Africa, for example, are less ready to stray from the mother in their playful exploration and more ready to run to her than are Western babies.[13] In a study of the outdoor behavior of English children, one and a half to two and a half years old, Anderson notes that they seldom stray more than 200 feet from their mothers. Characteristically the child moves in short bouts of no more than a few seconds. He stops between bouts for similar brief periods. Most of his walking time is spent drawing nearer to or farther from the mother. Objects and events in the environment do not appear to affect the way the child moves. The child does not necessarily move away from the mother because he is attracted by an object nor return to her in flight from an object. The movements have the playful character of experimentation. The child "moves a short distance from the mother, stops to look around, fixates the sources of sounds and visual stimuli and, in some cases, attracts the mother's attention to them. Intermingled with this scanning of the remote is an examination of the ground: he handles leaves, grass, stones and refuse; crawls or jumps backwards and forwards over verges, and attempts the shaking or climbing of obstacles."[14] Pointing is a common gesture. Any remote sight or sound that catches the child's attention is sufficient to elicit it. Often the adult is unable to discern the source of the stimulus. It may be imaginary. "A child will point to a part of the horizon where nothing is moving and tell the mother that a man is coming."[15] Of special interest in these observations is the child's apparent concern with the remote and the proximate. He points to the horizon and plays with stones at his feet, but he shows little interest in the middle ground.

Infants and young children tend to articulate a world in polarized categories. Things are noted and classified on the ground of maximal contrast. Language itself begins when the infant stops babbling indiscriminately and experiments with highly differentiated sounds. The first vowel is the wide open "a" and the first consonant the restricted "p" or "b" made with the lips. The first consonantal opposition is between nasal and oral stops (mama/papa); next comes the opposition of labials and dentals (papa/tata and mama/nana). Together these comprise the minimal consonantal system for all the languages of the world.[16] Between the sixth and the eighth month, we have noted, the infant begins to divide people into "familiar" and "strange." Shortly after he discriminates among inanimate toys. When toys are placed before him he grabs the one he favors rather than the one closest to him. A one-year-old child, held in the lap, raises his arms to gesture "up"; he wriggles and looks down when he means "down." Spatial opposites are clearly distinguished by a child two to two and a half years old. They include up and down, here and there, far and near, top and bottom, on and under, head and tail, front and back, front door and back door, front buttons and rear buttons, home and outside.[17] A toddler is able to verbalize some of these distinctions. They are not very specific. A young child distinguishes between "home" and "outside" as his play areas rather than "my bedroom" and "garden." The polar extremes are not understood equally well; for example, "here" has greater significance than "there," and "up" is more readily conceived than "down."[18]

The works of Piaget and his colleagues have repeatedly shown that sensorimotor intelligence precedes conceptual grasp, sometimes by several years. In the course of day-to-day activities a child displays spatial skills that are far beyond his intellectual comprehension. An infant six months old can discriminate between a square and a triangle, but the concept of square as a certain shape does not appear until a child is about four years old, when he can also draw it. Again, a young child may have the notion of the straight line as the trajectory of a moving object (the truck he pushes along the edge of the table), but the geometric concept of the straight line does not appear until the age of six or seven.[19] Prior to that age the child does not spontaneously draw a straight line and fails to grasp the idea of the diagonal.[20] A child beginning to walk soon walks to a purpose: he starts from a home base, heads toward the

Space, Place, and
the Child

object of desire, and returns to the starting point by a different
route. A vigorous child of three or four knows his way about
the house and, from time to time, visits the neighbors. These
sensorimotor achievements, however, do not imply a conceptual knowledge of spatial relations. Swiss children, five to six
years old, can go to school and return home by themselves.
They have difficulty explaining how this is done. One child
"remembers only where he starts and where he finishes and
that he has to go round a corner on the way. He cannot recall a
single landmark, and the journey he draws bears no relation to
his plan of the school and the surrounding district." Another
child "remembers names of roads, but not their order or the
places where he has to turn. His drawing is just an arc with a
number of points put in haphazardly to correspond with names
he can remember."[21]

A child's spatial frame of reference is restricted. Children's
art provides abundant hints of this restriction. For example, in
the child's drawing the level of water in a tilted glass tube is
shown at right angles to the sides of the tube, rather than as
parallel to the surface of the table that provides the horizontal
base line for the picture. Or, when a child is asked to draw a
chimney on the sloping roof of the house, he may place the
chimney at right angles to the sloping roof rather than to the
flat ground on which the house rests.[22] "Separation" is another
type of evidence that hints at the child's inability to depict, or
simply indifference to, the spatial relations among objects. For
example, the picture of the cowboy on his horse may show a
prominent gap between the cowboy's hat and his head, and
another gap between the cowboy and the horse.[23] Errors of this
kind suggest that the young child is more concerned with
things themselves—the water in the tube, the cowboy, and the
horse, than with their precise spatial relations. Parents know
how easily their young offspring get lost in an unfamiliar environment. Adults have acquired the habit of taking mental note
of where things are and of how to go from one place to
another. Children, on the other hand, are caught up in the
excitement of people, things, and events; going from one
place to another is not their responsibility.

Space, Place, and
the Child

Human beings live on the ground and see trees and houses from the side. The bird's-eye view is not ours, unless we climb a tall mountain or fly in an airplane. Young children rarely have the opportunity to assume a bird's-eye view of landscape. They are small people in a world of giants and of gigantic things not made to their scale. Yet children five or six years old show remarkable understanding of how landscapes look from above. They can read black-and-white vertical aerial photographs of settlements and fields with unexpected accuracy and confidence. They can pick out the houses, roads, and trees on aerial photographs even though these features appear greatly reduced in scale and are viewed from an angle and position unknown to them in actual experience. City children may have benefited from looking at pictures in magazines and television, but country children unexposed to these media are also good at interpreting vertical photographs of their environment.[24]

Perhaps one reason why young children can accomplish these feats of extrapolation is that they have played with toys. Although children are midgets in the world of adults, they are giants in their own world of toys. They look at toy houses and trains from a height and command their fates like Olympian gods. Susan Isaacs reports on a group of precocious English children who quickly learned about spatial relations through imaginative play.

The children had taken to modelling in plasticine whole scenes of places they had been to, such as the bathing-pool on the river, with the people in it. One day whilst they were modelling some such subject, an aeroplane passed over the garden, as often happened. The children all watched it, and shouted up to it as they usually did, "Come down, come down!". . . . [One child] said, "Perhaps he can see us?" And another, "I wonder what he sees, what we look like." I then suggested, "Perhaps we could make a model of the garden as it looks to the man in the plane?" This suggestion delighted them. We began on it at once and put several days' work into it. Some of the children climbed "as high up the ladder as we can get, to see how it looks from the plane." One boy of four-and-a-half realised spontaneously that from the plane only the tops of their own heads would be seen, and he dotted a number of small flat ovals over the paths of the model, "That's the children running about."[25]

Space, Place, and
the Child

In the period from 1950 to 1970, the ability of children of nursery-school age to understand aerial photographs has improved. Viewing aerial scenes on television and playing with simple constructional toys may have helped this progressive trend. On the other hand, over the same period children show no sign of greater sophistication in understanding viewpoints from opposite sides of a room or field.[26] It is easier for both the child and the adult to imagine how a pilot in his airplane sees the landscape than how a farmer on the opposite side of the hill sees it. We more readily assume a God-like position, looking at the earth from above, than from the perspective of another mortal living on the same level as ourselves. Moreover comprehension of environment suffers less after a 90-degree rotation of perspective from the horizontal than after a rotation of 40 to 50 degrees. The oblique view can be more difficult to interpret than the vertical view.

To the child, the picture taken from the side or at a small angle above ground has one major advantage over the map or aerial photograph: it is a more direct appeal to imaginative action. A child three and a half years old is already able to project himself kinesthetically into the illustration of his book. He looks at a picture and in his imagination he walks the path to the house and worms his way through its tiny door.[27] Central perspective creates an illusion of time and movement in a scene: the converging borders of a road that disappears into the door of a distant house are strong cues to action. In contrast, the vertical photograph invites the understanding of spatial relationships. The child is not prompted to initiate imaginative action—unless it is to drop bombs on the school house. A perspectival picture of the kind that is found in a storybook encourages an egocentric viewpoint: the child sees himself as the hero of the stage and is unable, or unwilling, to imagine how another actor—the little boy at the end of the road, for instance—would see him as he approaches. An aerial photograph or map, on the other hand, promotes an objective viewpoint. An objective viewpoint discourages action, especially those precipitous and self-dramatizing ventures that come naturally to the child.

Space, Place, and
the Child

How does a young child understand place? If we define place broadly as a focus of value, of nurture and support, then the mother is the child's primary place. Mother may well be the first enduring and independent object in the infant's world of fleeting impressions. Later she is recognized by the child as his essential shelter and dependable source of physical and psychological comfort. A man leaves his home or hometown to explore the world; a toddler leaves his mother's side to explore the world. Places stay put. Their image is one of stability and permanence. The mother is mobile, but to the child she nonetheless stands for stability and permanence. She is nearly always around when needed. A strange world holds little fear for the young child provided his mother is nearby, for she is his familiar environment and haven. A child is adrift—placeless— without the supportive parent.[28]

As the child grows he becomes attached to objects other than significant persons and, eventually, to localities. Place, to the child, is a large and somewhat immobile type of object. At first large things have less meaning for him than small ones because, unlike portable toys or security blankets, they cannot be handled and moved easily; they may not be available for comfort and support at moments of crisis. Moreover the child may develop ambivalent feelings toward certain places—large objects—that are his. For example, the high chair is his place. He is fed there and feeding is a source of satisfaction, but he is also fed things he doesn't like and he is imprisoned in his high chair. A child may view his crib with ambivalence. The crib is his cozy little world, but almost every night he goes to it with reluctance; he needs sleep but fears darkness and being left alone.

As soon as the child is able to speak with some fluency he wants to know the names of things. Things are not quite real until they acquire names and can be classified in some way. Curiosity about places is part of a general curiosity about things, part of the need to label experiences so that they have a greater degree of permanence and fit into some conceptual scheme. According to Gesell, at two or two and a half years the child comprehends "where." He has no clear image of the

Space, Place, and
the Child

intervening space between here and there, but he acquires a sense of place and of security when his "where?" is answered with "home," "office," or "big building." A year or so later, the child shows a new interest in landmarks. He recognizes and anticipates them when he is out for a walk or ride. Egocentrism is manifest in a tendency to think that all cars going in his direction must be going to his own place. The child also learns to associate persons with specific places. He is bewildered when he meets his nursery-school teacher downtown, because she seems to him dislocated; she upsets his system of classification.[29]

A child's idea of place becomes more specific and geographical as he grows. To the question, where do you like to play? a two-year-old will probably say "home" or "outdoors." An older child will answer "in my room" or "in the yard." Locations become more precise. "Here" and "there" are augmented by "right here" and "right there." Interest in distant places and awareness of relative distance increase. Thus a child three to four years old begins to use such expressions as "far away" and "way down" or "way off." To the question, where do you live? a two-year-old will probably say "home." A year or so later he may give the street name or even the name of the town, though infrequently.[30]

In elementary school years, how does a child's awareness of place deepen and expand? A study of first- and sixth-grade pupils in two midwestern American communities is suggestive.[31] The children are shown pictures of four types of places that are a part of their larger environment: village, city, farm, and factory. Of each place the question is asked: "What story does this picture tell?" The replies show marked individual differences. In general, those of the older children are much more sophisticated. Village, city, farm, and factory are familiar categories of place to sixth graders; they describe them with an assurance and facility comparable to those of adults. When shown a picture the older pupil is often able not only to say what it is (village, city, etc.), what it consists of, but also to put the place in its larger geographical context; he not only describes what the people shown in the picture are doing (mow-ing the lawn, shopping, etc.) but also attempts to explain how the place functions. In comparison, the first-grade pupil, when he looks at the picture of the village, is more likely to ignore its broader spatial setting; he may not even recognize it as a village, his attention being focused on its parts—the church, the school, the shop, and the road. The younger child tends to have little to say about the social and economic significance of the things he notices in the picture. Indeed the first grader's primary interest seems not to be the physical environment but the people in it, what the man or the little girl is doing. In general the first grader is less enthusiastic about places than the older child.

The geographical horizon of a child expands as he grows, but not necessarily step by step toward the larger scale. His interest and knowledge focus first on the small local community, then the city, skipping the neighborhood; and from the city his interest may jump to the nation and foreign places, skipping the region. At age five or six a child is capable of curiosity about the geography of remote places. How can he appreciate exotic locales of which he has no direct experience? Learning theory has yet to explain satisfactorily these apparent leaps in comprehension. It is not surprising, however, that a child can enjoy news of distant places, for he leads a rich life of fantasy and is at home in fantasyland before adults require him to dwell imaginatively in the real countries of a geography book. To an intelligent and lively child, experience is active searching and occasional wild extrapolations beyond the given: he is not bound by what he sees and feels in his home and local neighborhood.

What is the character of a young child's emotional tie to place? American first graders may recognize village, city corner, and farm as entities, but we have noted that the young pupils have less to say and are less enthusiastic about such places than is the case with older children. Except for nurseries and playgrounds few public places are made to the scale of young children. Do they feel a need to be in places that conform to their own size? Hints of such need exist. Infants, for example, are known to crawl under the grand piano, where they sit in an apparent state of bliss. Older children in their play seek out nooks and corners both in man-made environments and in nature. Spending the night in a tent or in a tree house at the backyard is a real treat and as much fun as being taken on a long trip to a real hunting lodge.

Feeling for place is influenced by knowledge, by knowing such basic facts as whether the place is natural or man-made and whether it is relatively large or small. A child five or six years old lacks this kind of knowledge. He may talk excitedly about the city of Geneva and Lake Geneva, but his appreciation of these places is certain to differ radically from that of an informed adult. He is at an age when he is likely to assume that both the city and the lake are artificial. He is also likely to assume they are comparable in size.[32]

Children, at least those of the Western world, develop a strong sense of property. They become strongly possessive. A child declares that certain toys are his, that the chair next to the mother's chair is his place, and he is not slow to defend what he considers to belong to him. Much of the child's combative possessiveness, however, is not evidence of genuine attachment. It arises out of a need for assurance of his own worth and for a sense of status among peers. An object or a corner of the room, valueless to the child one moment, suddenly becomes valuable when another child threatens to take possession. Once the first child has regained indisputable control, his interest in the toy or place quickly wanes.[33] This is not to deny that people, young and old, feel a need to anchor their personality in objects and places. All human beings appear to have personal belongings and perhaps all have need of a personal place, whether this be a particular chair in a room or a particular corner in a moving carriage.

Robert Coles believes that in the United States the children of migrant farm workers suffer because, among other reasons, they have no place that they can identify as their own over a period of time. Peter, for example, is a seven-year-old boy who travels up and down the East coast with his working parents. They seldom stay long in any farm. Peter helps to pick fruits and vegetables. He goes to school when he can. Coles writes:

Space, Place, and
the Child

To a boy like Peter a school building, even an old and not very attrac-
tively furnished one, is a new world—of large windows and solid
floors and doors and plastered ceilings and walls with pictures on
them, and a seat that one has, that one is given, that one is supposed
to own, or virtually own, for day after day, almost as a right of some
sort. After his first week in the first grade Peter said this: "They told
me I could sit in that chair and they said the desk, it was for me, and
that every day I should come to the same place, to the chair she said
was mine for as long as I'm there in that school—that's what they say,
the teachers, anyway."[34]

   Place can acquire deep meaning for the adult through the
steady accretion of sentiment over the years. Every piece of
heirloom furniture, or even a stain on the wall, tells a story. The
child not only has a short past, but his eyes more than the
adult's are on the present and the immediate future. His vitality
for doing things and exploring space is not suited to the reflective pause and backward glance that make places seem satu-
rated with significance. The child's imagination is of a special
kind. It is tied to activity. A child will ride a stick as though it
were a real horse, and defend an upturned chair as though it
were a real castle. In reading a book or looking at its pictures
he quickly enters a fantasy world of adventure. But a broken
mirror or an abandoned tricycle has no message of sadness.
And children are baffled when they are asked to interpret the
mood of a landscape or landscape painting. People have
moods; how can a scene or place look happy or sad?[35] Yet
adults, particularly educated adults, have no difficulty associating inanimate objects with moods. Young children, so imaginative in their own spheres of action, may look matter-of-factly
on places that to adults are haunted by memories.

4

Body,
Personal Relations,
and
Spatial Values

"Space" is an abstract term for a complex set of ideas. People of different cultures differ in how they divide up their world, assign values to its parts, and measure them. Ways of dividing up space vary enormously in intricacy and sophistication, as do techniques of judging size and distance. Nonetheless certain cross-cultural similarities exist, and they rest ultimately on the fact that man is the measure of all things. This is to say, if we look for fundamental principles of spatial organization we find them in two kinds of facts: the posture and structure of the human body, and the relations (whether close or distant) between human beings. Man, out of his intimate experience with his body and with other people, organizes space so that it conforms with and caters to his biological needs and social relations.

The word "body" immediately calls to mind an object rather than an animated and animating being. The body is an "it," and it is in space or takes up space. In contrast, when we use the terms "man" and "world," we do not merely think of man as an object in the world, occupying a small part of its space, but also of man as inhabiting the world, commanding and creating it. In fact the single term "world" contains and conjoins man and his environment, for its etymological root "wer" means

Body, Personal Relations,
and Spatial Values

man. Man and world denote complex ideas. At this point, we also need to look at simpler ideas abstracted from man and world, namely, body and space, remembering however that the one not only occupies the other but commands and orders it through intention. Body is "lived body" and space is humanly construed space.

Among mammals the human body is unique in that it easily maintains an upright position. Upright, man is ready to act. Space opens out before him and is immediately differentiable into front-back and right-left axes in conformity with the structure of his body. Vertical-horizontal, top-bottom, front-back and right-left are positions and coordinates of the body that are extrapolated onto space (Fig. 2). In deep sleep man continues to be influenced by his environment but loses his world; he is a body occupying space. Awake and upright he regains his

UPRIGHT HUMAN BODY, SPACE AND TIME

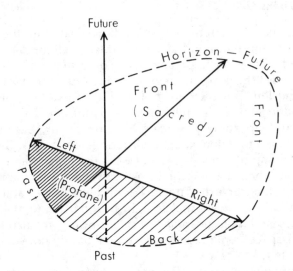

Figure 2. Upright human body, space and time. Space projected from the body is biased toward the front and right. The future is ahead and "up." The past is behind and "below."

Body, Personal Relations,
and Spatial Values

world, and space is articulated in accordance with his corporeal schema. What does it mean to be in command of space,
to feel at home in it? It means that the objective reference
points in space, such as landmarks and the cardinal positions,
conform with the intention and the coordinates of the human
body. Kant wrote in 1768:

Even our judgments about the cosmic regions are subordinated to the
concept we have of regions in general, insofar as they are determined
in relation to the sides of the body. . . . However well I know the
order of the cardinal points, I can determine regions according to that
order only insofar as I know towards which hand this order proceeds;
and the most complete chart of the heavens, however perfectly I
might carry the plan in my mind, would not teach me, from a known
region, North say, on which side to look for sunrise, unless, in addition to the positions of the stars in relation to one another, this region
were also determined through the position of the plan relatively to my
hands. Similarly, our geographical knowledge, and even our com-
monest knowledge of the position of places, would be of no aid to us
if we could not, by reference to the sides of our bodies, assign to
regions the things so ordered and the whole system of mutually relative positions.[1]

What does it mean to be lost? I follow a path into the forest,
stray from the path, and all of a sudden feel completely disoriented. Space is still organized in conformity with the sides
of my body. There are the regions to my front and back, to my
right and left, but they are not geared to external reference
points and hence are quite useless. Front and back regions
suddenly feel arbitrary, since I have no better reason to go
forward than to go back. Let a flickering light appear behind a
distant clump of trees. I remain lost in the sense that I still do
not know where I am in the forest, but space has dramatically
regained its structure. The flickering light has established a
goal. As I move toward that goal, front and back, right and left,
have resumed their meaning: I stride forward, am glad to have
left dark space behind, and make sure that I do not veer to the
right or left.

The human being, by his mere presence, imposes a schema
on space. Most of the time he is not aware of it. He notes its
absence when he is lost. He marks its presence on those ritual

Body, Personal Relations,
and Spatial Values

occasions that lift life above the ordinary and so force him to an awareness of life's values, including those manifest in space. Cultures differ greatly in the elaboration of spatial schemata. In some cultures they are rudimentary; in others they can become a many-splendored frame that integrates nearly all departments of life. Yet, despite the large outward differences, the vocabularies of spatial organization and value have certain common terms. These common terms are ultimately derived from the structure and values of the human body.

Upright and prone: these positions yield two contrary worlds. When a six-month-old infant sits up, Gesell and Amatruda report, "his eyes widen, pulse strengthens, breathing quickens and he smiles." For the infant the move from the supine horizontal to the seated perpendicular is already "more than a postural triumph. It is a widening horizon, a new social orientation."[2] This postural triumph and the consequent widening of horizon are repeated daily throughout a person's life. Each day we defy gravity and other natural forces to create and sustain an orderly human world; at night we give in to these forces and take leave of the world we have created. The standing posture is assertive, solemn, and aloof. The prone position is submissive, signifying the acceptance of our biological condition. A person assumes his full human stature when he is upright. The word "stand" is the root for a large cluster of related words which include "status," "stature," "statute," "estate," and "institute." They all imply achievement and order.[3]

"High" and "low," the two poles of the vertical axis, are strongly charged words in most languages. Whatever is superior or excellent is elevated, associated with a sense of physical height. Indeed "superior" is derived from a Latin word meaning "higher." "Excel" (*celsus*) is another Latin word for "high." The Sanskrit *brahman* is derived from a term meaning "height." "Degree," in its literal sense, is a step by which one moves up and down in space. Social status is designated "high" or "low" rather than "great" or "small." God dwells in heaven. In both the Old and the New Testament God was sometimes identified with heaven. Edwyn Bevan wrote: "The

Body, Personal Relations,
and Spatial Values

idea which regards the sky as the abode of the Supreme Being, or as identical with him, is as universal among mankind as any religious belief can be."[4]

In architecture, important buildings are put on platforms, and where the necessary technical skill exists they also tend to be the taller buildings. Of monuments this is perhaps invariably true: a tall pyramid or victory column commands greater esteem than a shorter one. Many exceptions to this rule occur in domestic architecture. The reason is clear: the symbolic advantages of the upper stories of a house can easily be outweighed by their practical problems. Before adequate piping systems were fitted into houses, water had to be carried up and wastes brought down by hand. Living in the upper stories required much work. Not only in ancient Rome but also in nineteenth-century Paris, the prestigious floor was the one above the ground-level shops. In tenements bordering the Champs-Elysées, the higher the rooms were, the poorer were the occupants: servants and poor artists occupied the garrets. In modern high-rise building, however, the handicap of vertical distance is overcome by sophisticated machinery, with the result that the prestige of elevation can reassert itself.

Residential locations show a similar hierarchy of values. As in a house the working parts lie concealed in the basement, so in a city the industrial and commercial base hugs the water's edge; and private homes rise in prestige with elevation.[5] The rich and powerful not only own more real estate than the less privileged, they also command more visual space. Their status is made evident to outsiders by the superior location of their residence; and from their residence the rich are reassured of their position in life each time they look out the window and see the world at their feet. Again, there are exceptions. A well-known one is Rio de Janeiro, where luxury high-rise buildings seek the convenience and attraction of the beach while the huts of the indigent cling to the steep slopes of the hills.

The prestige of the center is well established. People everywhere tend to regard their own homeland as the "middle place," or the center of the world.[6] Among some people there is also the belief, quite unsupported by geography, that they

Body, Personal Relations,
and Spatial Values

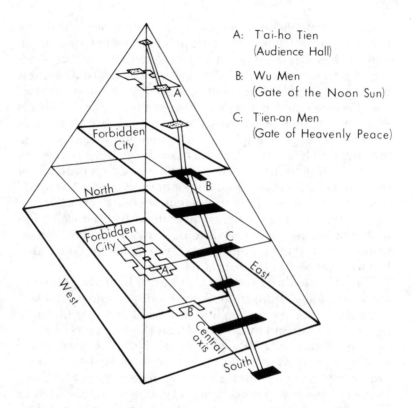

A: T'ai-ho Tien
(Audience Hall)

B: Wu Men
(Gate of the Noon Sun)

C: T'ien-an Men
(Gate of Heavenly Peace)

NORTHERN CITY OF PEKING

Figure 3. "Center" implies "elevation," and vice versa: the example of the northern city of Peking. The length of the southern avenue (central axis) should be read as height. "No matter how the natural terrain of China is formed, one always goes *up* to Peking" (N. Wu.). Reprinted with permission from Nelson I. Wu, *Chinese and Indian Architecture* (New York: George Braziller, 1963), Figure 136 "Plan of Peking interpreted as volume."

live at the top of the world, or that their sacred place is at the earth's summit (Fig. 3). The nomadic tribes of Mongolia, for example, once held the idea that they inhabit the top of a broad mound, the slopes of which are occupied by other races.[7] A common belief in Rabbinical literature is that the land

Body, Personal Relations,
and Spatial Values

of Israel stands higher above sea level than any other land, and that the Temple hill is the highest point in Israel.[8] Islamic tradition teaches that the most sacred sanctuary, the Kaaba, is not only the center and the navel of the world but also its highest point. Kaaba's spatial position corresponds to the polar star: "no place on earth is closer to heaven than Mecca."[9] This is why prayers said in its sanctuary are more clearly heard. When the explicit religious symbolism of center and height is weak, the physical elevation of the land nevertheless retains a certain prestige. Modern nations like to think that a high peak, if not the world's highest, lies within their borders. Lack of accurate measurement allows the imagination, fueled by patriotic fervor, to run wild. Even in the eighteenth century, educated Britons could consider Ben Nevis to be one of the loftiest mountains on earth.[10] India, Nepal, and China would no doubt each like to claim Mount Everest for its own.

In addition to the vertical-horizontal and high-low polarities, the shape and posture of the human body define its ambient space as front-back and right-left. Frontal space is primarily visual. It is vivid and much larger than the rear space that we can experience only through nonvisual cues. Frontal space is "illuminated" because it can be seen; back space is "dark," even when the sun shines, simply because it cannot be seen. The belief that eyes project light rays goes back at least to Plato (*Timaeus*) and persists to the Middle Ages and beyond. Another common feeling is that one's shadow falls behind the body even though in actual fact it often stretches to the front. On a temporal plane, frontal space is perceived as future, rear space as past. The front signifies dignity. The human face commands respect, even awe. Lesser beings approach the great with their eyes lowered, avoiding the awesome visage. The rear is profane (Fig. 2). Lesser beings hover behind (and in the shadow of) their superiors. In traditional China the ruler stands facing south and receives the full rays of the noon sun; he thus assimilates the male and luminous principle of *yang*. It follows from this that the front of the body is also yang. Inversely, the back of the ruler and the area behind him are *yin*, feminine, dark, and profane.[11]

Body, Personal Relations,
and Spatial Values

Every person is at the center of his world, and circumambient space is differentiated in accordance with the schema of his body. As he moves and turns, so do the regions front-back and right-left around him. But objective space also takes on these somatic values. Rooms at one end of the scale and cities at the other often show front and back sides. In large and stratified societies spatial hierarchies can be vividly articulated by architectural means such as plan, design, and type of decoration.

Consider some of the ways that front and back areas are distinguished in the Western world. Rooms are asymmetrically furnished: their geometrical center is not usually the focal point of interior space. For example, the focal point of the parlor may be the hearth, which is located at one end of the room. A typical lecture hall is sharply divided into front and back by the position of the lectern and the placement of chairs. Relation to other rooms rather than how furniture within a room is arranged may impart a bias to interior space: thus a bedroom has a front and back despite the symmetrical disposition of furniture, windows, and doors simply because one door opens out to the sitting room and another closes on the bathroom. Many buildings have clearly demarcated front and back regions. People may work in the same building and yet experience different worlds because their unequal status propels them into different circulatory routes and work areas. Maintenance men and janitors enter through service doors at the back and move along the "guts" of the building, while executives and their secretaries enter by the front door and move through the spacious lobby and well-lit passageways to their brightly furnished offices.[12] A middle-class residence typically presents an attractive front to impress and welcome social adults, and an unprepossessing rear for the use of people of low status such as delivery men and children.

Do cities have front and back regions? In the traditional Chinese city, front and back were clearly distinguished: there can be no mistaking the front and south, with its broad ceremonial avenue, for the back and north, which was reserved (at least in planning theory) for profane commercial use.[13] In the Near East and Europe, the distinctions were less systematically

Body, Personal Relations,
and Spatial Values

expressed in urban design. However, ancient walled cities boasted processional routes for use in royal and triumphal occasions; these routes probably had imposing front entrances. In the late Middle Ages and during the Renaissance, urban centers of political and ecclesiastical importance constructed magnificent front portals over walls that no longer served any military purpose. The monumentality of the portal symbolized the power of the ruler. It also functioned as an ideogram for the entire city, presenting a front that was meant to impress visitors and foreign potentates.[14]

The modern economic city has no planned front and back; it boasts no processional route or ceremonial gate, and its boundary is often arbitrary, marked by an inconspicuous signpost giving, as in the United States, the name and population of the borough. Yet the sense of "front" and "back" is not entirely absent. The width and appearance of the highway (landscaped or lined with giant posters) tell the motorist that he is entering the city by the front door.

If a modern city gives the impression of having a front and a back, that impression is as much the result of the direction and volume of traffic flow as of architectural symbols. On a still broader canvas, note how the historical movement of a people can give a sense of spatial asymmetry to a whole region or nation. St. Louis is the preeminent gateway to the West. The city has erected a soaring arch to dramatize its role as the front entrance to the Great Plains and beyond. Most people in the United States probably regard the northeastern seaboard as the nation's front. The nation's history is perceived to begin there. New York, in particular, has come to mean the front portal. Among the city's numerous nicknames, one is the Front Office of American Business. But more important than size and business power, New York owes its gateway image to the fact that through it so many immigrants entered the land of promise.

People do not mistake prone for upright, nor front for back, but the right and left sides of the body as well as the spaces extrapolated from them are easily confused. In our experience as mobile animals, front and back are primary, right and left are secondary. To move effectively we first rise and then go for-

Body, Personal Relations,
and Spatial Values

ward. The forward motion is periodically interrupted by the need to turn to the right or left. Suppose I am walking down the road and after a while make a turn to the right. An observer may say that I am now going to the right. But I do not feel that my direction is to the right in any absolute sense. I have made a turn to the right, but I continue to go forward to my goal. Right and left are distinctions I have to recognize. They are means, however, to my end which always lies in front.

The right and left sides of the human body are much alike in appearance and function. Some asymmetries exist: for example, the whorl of hair on the head turns to the right; the heart is slightly displaced to the body's left side; the two cerebral hemispheres are not equally well developed and have somewhat different functions; most people are right-handed, and when they move they show a tendency to veer to the right, perhaps as a result of a slight imbalance in vestibular control. Such small biological asymmetries do not seem adequate to explain the sharp distinctions in value attributed to the two sides of the body and to the spaces, social and cosmological, that extend from the body.

In nearly all the cultures for which information is available, the right side is regarded as far superior to the left. Evidence for the bias is particularly rich in Europe, the Middle East and Africa, but the bias is also well documented for India and Southeast Asia.[15] In essence, the right is perceived to signify sacred power, the principle of all effective activity, and the source of everything that is good and legitimate. The left is its antithesis; it signifies the profane, the impure, the ambivalent and the feeble, which is maleficent and to be dreaded. In social space the right side of the host is the place of honor. In cosmological space "the right represents what is high, the upperworld, the sky; while the left is connected with the underworld and the earth."[16] Christ, in pictures of the Last Judgment, has his right hand raised toward the bright region of Heaven, and his left hand pointing downward to dark Hell. A similar idea of the cosmos appears among the simple Toradja people of central Celebes. The right side is that of the living, a world of daylight; the left side is the dark underworld of the dead.[17] On a geographical plane, the ancient Arabs equated the left with the north and Syria. The word *šimâl* indicates both north and the left side. The Arabic word for Syria is Sam: its root meanings are "misfortune" or "ill augury" and "left." A related verb, *sa'ma*, means both to bring bad luck and to turn left. In contrast, the south and right side of the Arabic hearth is laden with blessings. The south is the flourishing land of Yemen, and its root *ymn* implies ideas of happiness and "right."[18] In west Africa the Temne regard east as the primary orientation. North is therefore to the left and considered dark; south, to the right, is light. To the Temne, thunder and lightning are prepared in the north, whereas "good breezes" come from the south.[19]

The Chinese view has special interest because it appears to be a striking exception to the rule. Like most people the Chinese are right-handed, but the honorable side for them is the left. In the great bipartite classification *yin* and *yang*, the left side is yang and belongs to the male, the right side is yin and belongs to the female. The basic reason is that the Chinese social and cosmological space centers on the ruler who mediates between heaven and earth. The ruler faces south and the sun. His left side is therefore east, the place of rising sun, and male (yang); his right side is west, the place of setting sun, and female (yin).[20]

Man is the measure. In a literal sense, the human body is the measure of direction, location, and distance. The ancient Egyptian word for "face" is the same as that for "south," and the word for "back of the head" carries the meaning of "north."[21] Many African and South Sea languages take their spatial prepositions directly from terms for parts of the body, such as "back" for "behind," "eye" for "in front of," "neck" for "above," and "stomach" for "within."[22] In the west African language, Ewe, the word for "head" stands for "peak" and the general spatial specifications of "over" and "above."[23] The principle of using nouns as prepositions expressing spatial relations can be extended beyond the body: for example, instead of "back," a word like "track" may be used to indicate "behind;" "under" may be designated by "ground" or "earth," and "over" by "air."[24]

Body, Personal Relations,
and Spatial Values

Spatial prepositions are necessarily anthropocentric, whether they are nouns derived from parts of the human body or not. As Merleau-Ponty put it: "When I say that an object is *on* a table, I always mentally put myself either in the table or in the object, and I apply to them a category which theoretically fits the relationship of my body to external objects. Stripped of this anthropological association, the word *on* is indistinguishable from the word *under* or the word *beside*."[25] Where is the book? It is on the desk. The answer is appropriate because it immediately helps us to locate the book by directing our attention to the large desk. It is hard to imagine a real-life circumstance in which the answer "the desk is *under* the book" is appropriate. We say an object is on, in, above, or under another object in reply to practical and even pressing concerns. Statements of location, therefore, normally give far more than simple locational facts. "I have locked my keys inside the car" tells where the keys are but it is also an anguished cry. "I am in my office" could mean, depending on the context, either "come in and see me," or "do not disturb." Only in the madhouse are statements like these purely locational; in the madhouse "the book is on the desk" and "the desk is under the book" are equivalent and interchangeable parts of speech.[26]

Folk measures of length are derived from parts of the body. Widely used are the breadth or length of finger or thumb; the span either from thumb to the tip of the little finger or to the tip of the forefinger; from the top of the middle finger to elbow (cubit), or over outstretched arms from finger tip to finger tip (fathom). Man-made objects in common use serve as ready measures of length, for example, the rod used to prod oxen, a spear, and customary segments of cord or chain. Estimates of longer distances draw on the experience and idea of effort. Thus the yard is a stride, the mile is one thousand paces, and the furlong (furrow long) is the stretch that the plow team can conveniently pull. Spear cast or bow shot gives rough units of distance, and even in the modern world we speak of "within a stone's throw," and "within shouting distance." Measures of capacity "include the hollow of the hand, the handful or arm-

Body, Personal Relations,
and Spatial Values

ful, the load of a man, beast, wagon or boat; the content of an egg, gourd, or other natural object; or of some manufactured object in common use, like a basket."[27] Measures of area are expressed in such units as an ox-hide, a mat, or cloak; the field that a yoke of oxen can plow in a day; and the land that can be sown with a given amount of seed.[28]

The human body and its subdivisions do not seem to provide common units for the estimation of area, as they do for the estimation of length and volume or capacity. Area is probably a more abstract concept than length or volume. Even in the simplest societies people must need to judge length and distance. "Capacity" is just as basic. The human body itself is a container. We know how it feels to be "full" or "empty." We experience directly the amount of food or water in our cupped hands or in our mouth. Adjectives for size, such as "big" and "small," apply primarily to volume and secondarily to area. The word "big" is in fact derived from Latin for "puffed cheek." Although in elementary geometry lessons we learn about area before we learn about volume, in common experience area is a sophisticated idea abstracted from the more primitive sense of capacity.

"Distance" connotes degrees of accessibility and also of concern. Human beings are interested in other people and in objects of importance to their livelihood. They want to know whether the significant others are far or near with respect to themselves and to each other. When a significant object is designated by a word or described in a phrase, the word or phrase suggests certain qualities in the object: "a fierce dog," "a broken spear," "a sick man." When we use these expressions, location and distance are implied although not explicitly given. "A fierce dog" is a dog too close to me for comfort, or tied to a post so that I am beyond its reach; "a broken spear" is the spear at hand but broken and hence useless. In Melanesian and in certain American Indian languages location and distance with respect to place or person are a necessary part of the description of objects. Codrington noted among Melanesians and Polynesians alike the habit of continually introducing adverbs of place and of direction such as up and down, hither

Body, Personal Relations,
and Spatial Values

and hence, seaward and landward. "Everything and everybody spoken of are viewed as coming or going, or in some relation to place, in a way which to the European is by no means accustomed or natural."[29] Of Kwakiutl, a Pacific coast Indian language, Boas wrote: "The rigidity with which location in relation to the speaker is expressed, both in nouns and in verbs, is one of the fundamental features of the language."[30] Various American Indian languages can express a thought such as "the man is sick" only by stating at the same time whether the subject of the statement is at a greater or lesser distance from the speaker or the listener and whether he is visible or invisible to them.[31]

Distance is distance from self (Fig. 4). In many languages, spatial demonstratives and personal pronouns are closely related so that it is difficult to say which class of words is earlier or later, which original or derivative. Words in both classes are half-mimetic, half-linguistic acts of indication. Personal pronouns, demonstrative pronouns, and adverbs of location closely implicate one another.[32] I am always *here*, and what is here I call *this*. In contrast with the here where I am, *you* are *there* and *he* is *yonder*. What is there or yonder I call *that*. "This" and "that" here perform the function of the German triple distinction "dies," "das," and "jenes." In non-European languages, a finely shaded range of demonstrative pronouns may be used to signify relative distances from self. Thus in Tlingit, an American Indian language, *he* indicates an object that is very near and always present; *ya* indicates an object very near and present, but a little farther off; *yu* indicates something so remote that it can be used as an impersonal article; *we* indicates a thing of far remoteness and is usually invisible.[33] The Chukchi in northeastern Siberia have as many as nine terms to express the position of an object in relation to the speaker.[34]

In English the demonstratives "this" and "that" are only a pair and so lack locational range; perhaps as a result their meanings become polarized and can carry high emotional charge. "We talked of this and that, but—alas—mostly that." The word "that" clearly suggests conversational topics both remote and trivial.[35] In *Richard II*, Shakespeare succeeds in

Body, Personal Relations, and Spatial Values

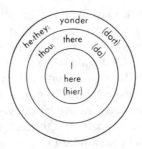

A. Personal pronouns and
spatial demonstratives

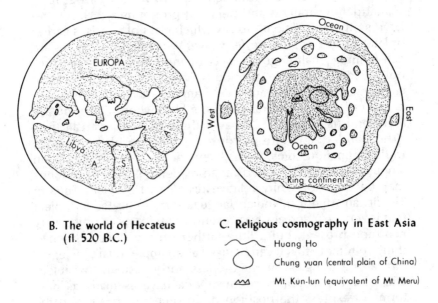

B. The world of Hecateus
(fl. 520 B.C.)

C. Religious cosmography in East Asia

Huang Ho

Chung yuan (central plain of China)

Mt. Kun-lun (equivalent of Mt. Meru)

Figure 4. Egocentric (A) and ethnocentric (B-G) organizations of space from ancient to modern times, in literate and nonliterate societies. Figure 4G is reprinted with permission from Torsten Hägerstrand, "Migration and area: survey of a sample of Swedish migration fields and hypothetical considerations on their genesis," *Lund Studies in Geography*, Series B, Human Geography, vol. 13, 1957, page 54.

Body, Personal Relations,
and Spatial Values

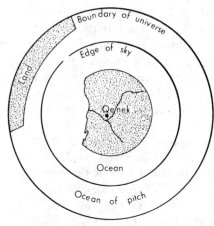

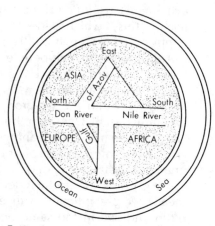

D. Yurok (California Indian) idea of the world

E. T—O map, after Isidore, Bishop of Seville (570—636 A.D.)

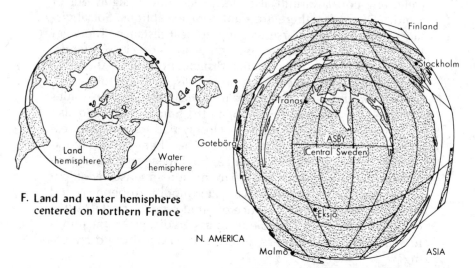

F. Land and water hemispheres centered on northern France

G. Map with azimuthal logarithmic distance scale, centered on central Sweden

Body, Personal Relations,
and Spatial Values

evoking a sense of patriotic fervor partly through the insistent use of "this," which is identified with "we the English." "This happy breed of men, this little world, this precious stone. . . ."

A distinction that all people recognize is between "us" and "them." We are *here*; we are *this* happy breed of men. They are *there*; they are not fully human and they live in *that* place. Members within the we-group are close to each other, and they are distant from members of the outside (they) group. Here we see how the meanings of "close" and "distant" are a compound of degrees of interpersonal intimacy and geographical distance. It may not be possible to decide which sense is primary and which derivative.[36] "We are close friends" means we are intimate with each other, we see each other often and live in the same neighborhood. Being close combines the two meanings of intimacy and geographical proximity. As the friend moves farther and farther away geographically, emotional warmth also declines: "out of sight, out of mind." Of course, there are numerous exceptions. Social distance may be the inverse of geographical distance. The valet lives close to his master but they are not close friends. Psychologically, absence (spatial distance) can make the heart grow fonder. Such exceptions do not disprove the rule.

We have indicated that certain spatial divisions and values owe their existence and meaning to the human body, and also that distance—a spatial term—is closely tied to terms expressive of interpersonal relationships. This theme is easily expanded. We may ask, for instance, how space and the experience of spaciousness are related to the human sense of competence and of freedom. If space is a symbol for openness and freedom, how will the presence of other people affect it? What concrete experiences enable us to assign distinctive meanings to space and spaciousness, to population density and crowding?

5

Spaciousness and Crowding

Space and spaciousness are closely related terms, as are population density and crowding; but ample space is not always experienced as spaciousness, and high density does not necessarily mean crowding. Spaciousness and crowding are antithetical feelings. The point at which one feeling turns into another depends on conditions that are hard to generalize. To understand how space and human number, spaciousness and crowding are related, we need to explore their meaning under specific conditions.[1]

Consider space. As a geometrical unit (area or volume), it is a measurable and unambiguous quantity. More loosely speaking, space means room; the German word for space is *raum*. Is there room for another crate of furniture in the warehouse? Is there room for another house on the estate? Does the college have room for more students? Although these questions have a similar grammatical form and all use the word "room" appropriately, the meaning of "room" differs in each case. The first question asks whether more objects can be put in, and the answer calls for simple and objective measurement. The second and third questions show that room can mean more than physical space; it suggests spaciousness. The question is not whether a house can be fitted physically into an estate, but whether the site is sufficiently spacious. And a college must have not only adequate classrooms and facilities, but it should feel commodious and liberating to students who go there to enlarge their minds.

Spaciousness is closely associated with the sense of being free. Freedom implies space; it means having the power and enough room in which to act. Being free has several levels of meaning. Fundamental is the ability to transcend the present condition, and this transcendence is most simply manifest as the elementary power to move. In the act of moving, space and its attributes are directly experienced. An immobile person will have difficulty mastering even primitive ideas of abstract space, for such ideas develop out of movement—out of the direct experiencing of space through movement.

An infant is unfree, and so are prisoners and the bedridden. They cannot, or have lost their ability to, move freely; they live in constricted spaces. An old person moves about with increasing difficulty. Space seems to close in on him. To an energetic child, a flight of stairs is a link between two floors, an invitation to run up and down; to an old man it is a barrier between two floors, a warning for him to stay put. The physically vital—children and athletes—enjoy a sense of spatial expansiveness little known to office-bound workers, who listen to tales of physical prowess with a mixture of admiration and envy. Eric Nesterinko, a hockey player with the Toronto Maple Leafs, described how even as a child winning was secondary to the joy of movement. "When I was a kid," he recalled, "to really move was my delight. I felt released because I could move around anybody. I was free." As a middle-aged man of thirty-eight Nesterinko retained his delight in spatial freedom. He recalled how on a cold, clear, and crisp afternoon, he saw a huge sheet of ice in the street. Unthinkingly he drove his car on the ice, got out and put on his skates. The hockey player said: "I took off my camel-hair coat. I was just in a suit jacket, on my skates. And I flew. Nobody was there. I was as free as a bird. . . . Incredible! It's beautiful! You're breaking the bounds of gravity. I have a feeling this is the innate desire of man."[2]

Tools and machines enlarge man's sense of space and spaciousness. Space that is measurable by the reach of one's outstretched arms is a small world compared with one that is measured by the distance of a spear throw or arrow shot. The body can feel both measures. Size is the way a person feels as he stretches his arms; it is the experience of the hunter as he *throws* his spear, *feels* it shoot out of his hand, and *sees* it disappear into the distance. A tool or machine enlarges a person's world when he feels it to be a direct extension of his corporeal powers. A bicycle enlarges the human sense of space, and likewise the sports car. They are machines at man's command. A perky sports car responds to the driver's slightest wish. It opens up a world of speed, air, and movement. Accelerating over a straight road or swerving over a curve, momentum and gravity—these dry terms out of a physics book—become the felt qualities of motion. Small aircrafts of the kind in use during the 1920s are capable of extending man's freedom, his space, as well as putting the human being into a more intimate relationship with the vastness of nature. The French writer and pilot Antoine de Saint-Exupéry expressed it this way:

The machine which at first blush seems a means of isolating man from the great problems of nature, actually plunges him more deeply into them. As for the peasant so for the pilot, dawn and twilight become events of consequence. His essential problems are set him by the mountain, the sea, the wind. Alone before the vast tribunal of the tempestuous sky, the pilot defends his mails [sic] and debates on terms of equality with those three elemental divinities.[3]

When the Paleolithic hunter drops his hand ax and picks up a bow and arrow, he takes a step forward in overcoming space and yet space expands before him: things once beyond his physical reach and mental horizon now form a part of his world. Imagine a man of our time who learns first to ride a bicycle, then to drive a sports car, and eventually to pilot a small aircraft. He makes successive gains in speed; greater and greater distances are overcome. He conquers space but does not nullify its sensible size; on the contrary, space continues to open out for him. When transportation is a passive experience, however, conquest of space can mean its diminishment. The speed that gives freedom to man causes him to lose a sense of spaciousness. Think of the jetliner. It crosses the continent in a few hours, yet its passengers' experience of speed and space is probably less vivid than that of a motorcyclist roaring down a freeway. Passengers have no control over the machine and cannot feel it as an extension of their organic powers. Passengers are luxury crates—safely belted in their seats—being transported passively from point to point.

Space is a common symbol of freedom in the Western world. Space lies open; it suggests the future and invites action. On the negative side, space and freedom are a threat. A root meaning of the word "bad" is "open." To be open and free is to be exposed and vulnerable. Open space has no trodden paths and signposts. It has no fixed pattern of established human meaning; it is like a blank sheet on which meaning may be imposed. Enclosed and humanized space is place. Compared to space, place is a calm center of established values. Human beings require both space and place. Human lives are a dialectical movement between shelter and venture, attachment and freedom. In open space one can become intensely aware of place; and in the solitude of a sheltered place the vastness of space beyond acquires a haunting presence. A healthy being welcomes constraint and freedom, the boundedness of place and the exposure of space. In contrast, the claustrophobe sees small tight places as oppressive containment, not as contained spaces where warm fellowship or meditation in solitude is possible. An agoraphobe dreads open spaces, which to him do not appeal as fields for potential action and for the enlargement of self; rather they threaten self's fragile integrity.[4]

Physical environment can influence a people's sense of size and spaciousness. On the small Melanesian island of Tikopia, which is only three miles long, the islanders have little conception of landmass size. They have wondered whether any land exists from which the sound of ocean waves cannot be heard.[5] China, in extreme contrast, stretches over a continent. Its people have learned to envisage vast distances and to think of them in dread, for they can mean the separation of friends and

Spaciousness and Crowding

lovers. Early Chinese literature used the expression, one thousand *li*, to evoke a sense of great distance. By the Han dynasty "ten thousand li" came into fashion. Poetic hyperbole required adjustment as geographical knowledge increased. Moreover, as geographical knowledge increased, poets could use contrasting natural environments to evoke a sense of distance and of separation. The following lines from a poem written during the Han period illustrate a sentiment and the method employed to heighten it:

> On and on, going on and on,
> away from you to live apart,
> ten thousand li and more between us,
> each at opposite ends of the sky.
> The road I travel is steep and long;
> who knows when we meet again?
> The Hu horse leans into the north wind;
> The Yueh bird nests in southern branches;
> day by day our parting grows more distant . . .[6]

Hu is a term for the area north of China extending from Korea to Tibet; *Yueh* designates the area around the mouth of the Yangtze River. Thus an abstract hyperbole of distance, ten thousand li, is fleshed out with the imagery of two specific regions and their contrasting ecologies.

Is the feeling of spaciousness identifiable with particular kinds of environment? A setting is spacious if it allows one to move freely. A room cluttered with furniture is not spacious whereas a bare hall or a public square is, and children let loose in either of them tend to respond by rushing about. A broad treeless plain looks open and expansive. The relationship of environment to feeling seems clear; but in fact, general rules are difficult to formulate. Two factors confuse the issue. One is that the feeling of spaciousness feeds on contrast. For example, a house is a compact and articulated world compared with the valley outside. From inside the house the valley beyond looks broad and lacking in definition, but the valley is itself a well-defined hollow compared with the plain onto which it in turn opens. The second factor is that culture and experience strongly influence the interpretation of environment. Americans have learned to accept the open plains of the West as a symbol of opportunity and freedom, but to the Russian peasants boundless space used to have the opposite meaning. It connoted despair rather than opportunity; it inhibited rather than encouraged action. It spoke of man's paltriness as against the immensity and indifference of nature. Immensity oppressed. Maxim Gorky wrote:

The boundless plain upon which the log-walled, thatch-roofed village huts stand huddled together has the poisonous property of desolating a man's soul and draining him of all desire for action. The peasant may go beyond the limits of his village, take a look at the emptiness all about him, and after a while he will feel as if this desolation had entered into his own soul. Nowhere are lasting traces of toil to be seen. . . . As far as the eye can see stretches an endless plain, and in the midst of it stands an insignificant wretched little man, cast away upon this dreary earth to labor like a galley slave. And the man is overwhelmed by a feeling of indifference which kills his capacity to think, to remember past experience, and to draw inspiration from it.[7]

The problem of how environment and feeling are related comes to a head with the question, can a sense of spaciousness be associated with the forest? From one viewpoint, the forest is a cluttered environment, the antithesis of open space. Distant views are nonexistent. A farmer has to cut down trees to create space for his farmstead and fields. Yet once the farm is established it becomes an ordered world of meaning—a place—and beyond it is the forest and space.[8] The forest, no less than the bare plain, is a trackless region of possibility. Trees that clutter up space from one viewpoint are, from another, the means by which a special awareness of space is created, for the trees stand one behind the other as far as the eyes can see, and they encourage the mind to extrapolate to infinity. The open plain, however large, comes visibly to an end at the horizon. The forest, although it may be small, appears boundless to one lost in its midst.

Whether forested mountains or grassy plains serve as image of spaciousness depends, at least in part, on the nature of a people's historical experience. In the period of vigorous European expansion in the nineteenth century migrants moved generally from forests to grasslands. The grasslands of North America at first

provoked dread; they lacked definition compared with the re-
ticulated spaces of the settled and forested East. Later, Americans
interpreted the plains more positively: the Eastern seaboard
might have finely ordered places but the West claimed space and
freedom. In contrast with the American experience, in China the
old centers of population were located in the relatively open
country of the subhumid and semiarid North. The movement of
the people was to the hilly and forested South. Could the Chinese
have associated the forested South with a sense of spaciousness?
At least some of them did. Poetry of the Han period, for example,
described the wildernesses of the South with awe; there, up-
rooted officials from the North encountered a vast and seemingly
primordial world of mist-wrapped mountains and lakes. It was in
South China that nature poetry and landscape painting reached
their highest development: both arts contrasted spacious nature, a world of shifting light and of peak behind mountain peak
fading into infinity, with the closed and formal world of man.[9]

Space is, of course, more than a complex and shifting viewpoint or feeling. It is a condition for biological survival. But the
question of how much space a man needs to live comfortably has
no simple answer. Space as resource is a cultural appraisal. In the
Orient a farm family can live contentedly on a few intensely
worked acres; in the United States, in 1862, a quarter section or
160 acres was judged the proper size for a yeoman's homestead.
Level of aspiration clearly affects one's sense of spatial adequacy.
Aspiration is culturally conditioned. Traditional China, for instance, had many small landlords who were content to live off
their rents and enjoy their leisure rather than work and invest
their income in enlarging their holdings. In capitalist Western
societies, aspiration and the entrepreneurial spirit have been and
are much stronger. To the truly acquisitive the goods that are
owned seldom seem quite enough. Space, fully satisfactory for
present operations, still may not feel sufficient. Biological appe-
tites soon reach their natural limits, but ultrabiological
yearning—which readily takes the perverted form of greed—is
potentially boundless. Tolstoy was led to ask, in exasperation,
"How much land does a man need?"—the title of an eloquent
fable in which he gave his answer. Although Tolstoy's question

sounded matter-of-fact, his answer and those offered by others usually veiled profound political and moral commitments.

Space is a resource that yields wealth and power when properly exploited. It is worldwide a symbol of prestige. The "big man" occupies and has access to more space than lesser beings. An aggressive ego endlessly demands more room in which to move. The thirst for power can be insatiable—especially power over money or territory, since financial and territorial growths are basically simple additive ideas that require little imaginative effort to conceive and extrapolate. The collective ego of a nation has made claims for more living space at the expense of its weaker neighbors; once a nation starts on the road of successful aggrandizement it could see no compelling limit to growth short of world dominion. For the aggressive nation as for the aggressive individual, the contentment that goes with the feeling of spaciousness is a mirage that recedes as one acquires more space.

Space, a biological necessity to all animals, is to human beings also a psychological need, a social perquisite, and even a spiritual attribute. Space and spaciousness carry different sets of meaning in different cultures. Consider the Hebraic tradition, one that has had a strong influence on Western values. In the Old Testament, words for spaciousness mean in one context physical size and in others psychological and spiritual qualities. As a physical measure spaciousness is "a good and a broad land, a land flowing with milk and honey" (Exodus, 3, 8). Israelites were concerned with the size of the promised land. They could not themselves take up arms and enlarge it at their neighbors' expense, but God could sanction their venture. "For I will cast out nations before you and enlarge your borders; neither shall any man desire your land" (Exodus, 34, 24). Psychologically, space in the Hebraic tradition means escape from danger and freedom from constraint. Victory is escape "into a broad place." "He brought me forth into a broad place; he delivers me, because he delighted in me" (Psalm 18, 19). In Psalm 119 the language of spaciousness is applied to the intellectual enlargement and spiritual freedom of the man who knows the Torah. "I will run in the way of thy commandments when those enlargest my understanding" (verse 32). On the spiritual plane, space connotes deliverance and salvation.[10]

Thus far we have explored the meaning of spaciousness without regard to the presence of other people. Solitude is a condition for acquiring a sense of immensity. Alone one's thoughts wander freely over space. In the presence of others they are pulled back by an awareness of other personalities who project their own worlds onto the same area. Fear of space often goes with fear of solitude. To be in the company of human beings— even with one other person—has the effect of curtailing space and its threat of openness. On the other hand, as people appear in space, for every one a point is reached when the feeling of spaciousness yields to its opposite—crowding. What constitutes crowding? We may say of a forest that it is crowded with trees and of a room that it is crowded with knick-knacks. But primarily people crowd us; people rather than things are likely to restrict our freedom and deprive us of space.

As an extreme example of how others can affect the scale of our world, imagine a shy man practising the piano in the corner of a large room. Someone enters to watch. Immediately the pianist feels spatial constraint. Even one more person can seem one too many. From being the sole subject in command over space, the pianist, under the gaze of another, becomes one object among many in the room. He senses a loss of power to order things in space from his unique perspective. Inanimate objects seldom produce this effect, although a man may feel ill at ease in a room full of ancestral portraits. Even a piece of furniture can seem to possess an obtrusive presence. Things, however, have this power only to the degree that people endow them with animate or human characteristics. Human beings possess this power naturally. But society can deprive them of it. Human beings can be treated as objects so that they are no more in one's way than are bookshelves. A rich man is surrounded by servants, yet they do not crowd him, for their low status makes them invisible—part of the woodwork.

Crowding is a condition known to all people at one time or another. People live in society. Whether one is an Eskimo or a New Yorker, occasions will arise when he has to work or live closely with others. Of the New Yorker this is obviously true, but even Eskimos do not always move on the broad open stage of the

Spaciousness and Crowding

Tundra; in the course of many dark and long nights they have to bear with each other's company in ill-ventilated huts. The Eskimo, though less often than the New Yorker, must on occasion screen the stimulus of other people by turning them into shadows and objects. Etiquette and rudeness are opposite means to the same end: helping people to avoid contact when such contact threatens to be too intense.

A sense of crowding can appear under highly varied conditions and at different scales. Two persons in one room, we have noted, can constitute a crowd. The pianist stops playing and leaves. Consider the large-scale phenomenon of crowding and migration. In the nineteenth century many Europeans abandoned their small farms, crowded dwellings, and polluted cities for the virgin lands of the New World. We rightly interpret the migration as motivated by the desire to seek opportunities in a freer and more spacious environment. Another major flow of people in both Europe and North America was from the countryside and small settlements to the large cities. We tend to forget that rural-urban migration, like the earlier movement across the ocean and into the New World, could also be motivated by the impulse to escape crowding. Why did country people, especially the young, leave their small hometowns for the metropolitan centers? One reason was that the hometown lacked room. The young considered it crowded in an economic sense because it did not provide enough jobs, and in a psychological sense because it imposed too many social constraints on behavior. The lack of opportunity in the economic sphere and of freedom in the social sphere made the world of the isolated rural settlement seem narrow and limited. Young people abandoned it for the jobs, the freedom, and—figuratively speaking—the open spaces of the city. The city was the place where the young believed they could move ahead and better themselves. Paradoxically the city seemed less "crowded" and "hemmed in" than the countryside of diminishing opportunities.

Crowding is an awareness that one is observed. In a small town people "watch out" for one another. "Watch out" has both the desirable sense of caring and the undesirable one of idle—and perhaps malicious—curiosity. Houses have eyes. Where they are built close together the neighbors' noises and the neighbors' concern constantly intrude. Where they are isolated privacy is better preserved—but not guaranteed; such is human ingenuity. On Shetland Isle, off the coast of Scotland, cottages are spaced far apart. Visual intrusion nonetheless persists. According to the sociologist Ervin Goffman, cottagers of seafaring background use pocket telescopes to observe their neighbor's activities. Distance notwithstanding the Shetlander can, from his own home, keep a neighborly eye on who is visiting whom.[11]

Trees or boulders may be dense in a wilderness area, but nature lovers do not see it as cluttered. Stars may speckle the night sky; such a sky is not viewed as oppressive. To city sophisticates nature, whatever its character, signifies openness and freedom. Human beings, if they are engaged in earning a livelihood from nature, blend into the natural scene and do not disrupt its solitude. In the densely peopled rice lands of the Orient the farmers rhythmically at work are barely visible to the outside observers, so much do they seem to belong to the earth. Is Java crowded? Its average population density of more than a thousand persons per square mile makes it one of the most crowded regions in the world. Yet here is the ambivalent response of the environmental psychiatrist Aristide Esser. While Esser recognizes the objective fact of Java's high population density, he says of the island in which he was born: "The beauty of its landscape and the relaxed, open mind of its people produced in me images of freedom on an unlimited beautiful world." In contrast, Holland—the country that gave Esser his schooling—looked "ridiculously petite" and "oppressive."[12]

Whether nature retains its air of solitude or not may bear little relation to the number of people living and working in it. Solitude is broken not so much by the number of organisms (human and nonhuman) in nature as by the sense of busy-ness—including the busy-ness of the mind—and of cross-purposes, actual and imagined. Mary McCarthy observed that an awareness of being "at one" with nature itself begins to constitute an intrusion: "Two fishermen pulling in a net on the seashore appear natural, but two poets brooding side by side on the same strand would be ridiculous—one solitude too many."[13]

Spaciousness and Crowding

People are social beings. We appreciate the company of our own kind. How physically close we tolerate or enjoy the presence of others, for how long, and under what conditions vary notice-ably from culture to culture. The Kaingáng Indians of the Amazon basin like to sleep in groups, locked limb to limb. They like to touch and fondle each other; they seek physical (nonsexual) intimacy for comfort and reassurance.[14] In another sparsely settled part of the world, the Kalahari desert, the !Kung Bushmen live under crowded conditions. Patricia Draper noted that in a Bushman camp the average space each person has is only 188 square feet, which is far less than the 350 square feet per person regarded as the desirable standard by the American Public Health Association. Space in a Bushman camp is arranged to ensure maximum contact. "Typically huts are so close that people sitting at different hearths can hand items back and forth without getting up. Often people sitting around various fires will carry on long discussions without raising their voices above normal conversa-tional levels."[15] The desert does not lack space. Bushmen live close by choice, and they do not show symptoms of biological stress.

In Western industrial society, working-class families are known to tolerate a much higher residential density than do middle-class families. And the reason is not simply because workers have little choice. Proximity to others is desired. Subur-ban retreats, each sitting on its own half-acre of lawn, are not necessarily the envy of working-class families accustomed to the bustle and color of an older neighborhood. Such families view the middle-class suburb with suspicion; it seems cold and ex-posed. Human proximity, human contact, and an almost con-stant background of human noises are tolerated, even wel-comed. In a new housing project in Chile, for example, the working-class residents shifted furniture from their living rooms into the hall so that they could be together, as was their custom. In England, it is true, a study of families who moved from old crowded dwellings to a new relatively spacious housing estate showed that the families benefited from the change; they were less tense because privacy was more readily available. On the other hand, at least for a time bedrooms were shared unnecessar-

Spaciousness and Crowding

ily, and by choice homework and other tasks were done in company.[16]

A crowd can be exhilarating. Young and old, from all levels of society, know this.[17] One group differs from another largely in the desired period of exposure, the occasion, and the preferred setting. What do English workers and their families do during their summer holidays? Large numbers flock to the beach. They escape their cramped quarters in industrialized cities for the milling crowds of Blackpool and Southend. The crowd at the seaside, far from being a nuisance, is a major attraction. What is a parade, a state fair, a charity bazaar, a revival meeting, or a football game without the multitudes?

Young Americans from well-to-do families are often strong partisans of nature and of the wilderness experience. At the same time they seem to like crowds. Protest marches against social injustice and war arise out of true indignation; yet the young marchers surely also enjoy the camaraderie, the sense of group solidarity in a righteous cause, and the sheer pleasure of swimming in a sea of their own kind.

Outdoor rock festivals capture the essential ambivalence of the young. On the one hand there is the outdoor setting, the bra-less freedom, and the nudity; on the other the enormous crowd—more densely packed than any street in Manhattan—and the blare of electronically amplified music. On July 28, 1973 some 600,000 youngsters attended the outdoor rock festival at Watkins Glen, New York. The fans were packed shoulder to shoulder on a 90-acre grassy knoll. "From the air," the New York Times reported, "the fence-enclosed concert site looked like a human ant hill surrounded by acres of meadows packed with cars and brightly colored tents. The knoll was so clogged with people that one nineteen-year-old girl from Patchogue, Long Island, reported that a trip to a portable toilet and back several hundred yards away had taken three hours."[18] When we consider the baking sun, the human swarm, the poor toilet facilities, and the high intake of beer and wine, it is reasonable to expect acute physical and mental stress, pent-up frustration, outbursts of anger, and fisticuffs. In fact the crowd was even-tempered and well behaved. The absence of serious inci-

Spaciousness and Crowding

dents surprised both local residents and the police. Music was clearly not the sole attraction of the festival—the crowd was its own entertainment.

People crowd us but they can also enlarge our world. Heart and mind expand in the presence of those we admire and love. When Boris Pasternak's heroine Lara enters a room, it is as if a window were flung open and the room filled with light and air.[19] When people work together for a common cause, one man does not deprive the other of space; rather he increases it for his colleague by giving him support. "The more angels there are, the more free space," said the erudite scientist-theologian Swedenborg (1688–1772), for the essence of the angel is not the use of space but its creation through selfless acts.[20] On the other hand people are a common cause of our frustration: their will thwarts ours. People often stand in our way, and when they do they are rarely presumed innocent, like tree stumps and furniture that cannot help being where they are. Inside a packed stadium other humans are welcome; they add to the excitement of the game. On the way home, driving along the clogged highway, other humans are a nuisance. When a car ahead is stalled, we feel almost as though the driver had intended mischief. The stadium has a higher density of people than the highway, but it is on the highway that we taste the unpleasantness of spatial constraint.

Conflicting activities generate a sense of crowding. In a small city apartment, a harassed mother tries to cook, feed the infant, scold the toddler who has spilled food on the floor, and answer the doorbell, all at the same time. A work-weary father returns home and cannot find a quiet corner to himself, away from his bumptious and loquacious children. If such a family were to move into adequate quarters, tension would no doubt decline and family contentment increase. However, human beings are so adaptable that under certain favorable conditions they can wring an advantage even from residential crowding—namely, a kind of indiscriminate, gregarious human warmth. Working-class people have sometimes achieved this warmth, as writers of working-class background—notably D. H. Lawrence and Richard Hoggart—have observed. In the congested home of an English worker's family, it is difficult to be alone, to think alone, or to read quietly. Not only things but people are shared. Mother is "our mom," father is "our dad," and the daughter is "our Alice." Hoggart captured the jumbled yet intensely human world of acceptance and sharing with this living room scenario:

There is the wireless or television, things being done in odd bouts, or intermittent snatches of talk . . . ; the iron thumps on the table, the dog scratches and yawns or the cat meaows to be let out; the son dries himself on the family towel near the fire, whistles or rustles the communal letter from his brother in the army which has been lying on the mantelpiece behind the photo of his sister's wedding; the little girl bursts into a whine because she is too tired to be up at all.[21]

Out of the crowded room a haven of warmth and tolerance is created. What is the loss? What is the cost of this successful adaptation to crowding? The cost appears to be a chance to develop deep inwardness in the human personality. Privacy and solitude are necessary for sustained reflection and a hard look at self, and through the understanding of self to the full appreciation of other personalities.[22] A man is not only a miner, he is not just "our dad," but also an individual with whom prolonged exchange—opening up worlds in sustained conversation or common enterprise—ought to be possible. Spatial privacy does not, of course, guarantee solitude; but it is a necessary condition. Living constantly in a small, close-knit group tends to curtail the enlargement of human sympathy in two antipodal directions: toward one pole, an intimacy between unique individuals that transcends camaraderie and kinship ties; and toward the other, a generalized concern for human welfare everywhere.[23]

The world feels spacious and friendly when it accommodates our desires, and cramped when it frustrates them. Frustration differs in seriousness. Among the affluent it may be no more than being tied up in traffic or having to wait for a particular table at a favorite restaurant. To the urban poor frustration often means shuffling in long and slow lines before the welfare or employment official's desk. There is also frustration at the most fundamental level: the awareness that land and resources are limited and too many stomachs remain unfilled. The mass of mankind has known this kind of deprivation, this sense of crowding. Thomas Malthus's insight into the relationship between resource and population commands special respect for its precision; the awareness of crowding in the Malthusian sense, however, has long been commonplace and widespread. It has existed in areas of both high and low population densities, in India, southeast Asia, and Europe as well as in the sparsely settled parts of North America.[24] Folklore and legends from such lands relate the theme of the overcrowded earth in varying detail and degree of explicitness. A Malthusian tale commonly begins with a world in which death was unknown. Human beings bred until the earth could no longer support them and there was great suffering. The story might then continue like this. God commanded an angel to kill people after they had reached a certain age. The angel demurred because he did not relish human curse; so God allowed the angel to hide his deeds behind a screen of diseases, accidents, and wars. Here is another Malthusian tale. According to the Iglulik Eskimos, death did not exist in the earliest times. The first humans lived on an island in Hudson Strait; they multiplied rapidly but none ever left home. Eventually so many people crowded the island that it could not support them and began to sink. An old woman shouted: "Let it be so ordered that human beings can die, for there will no longer be room for us on earth." And her wish was granted.[25]

Eskimos hunt in small groups over the broad open spaces of the Arctic coast. Urban crowding and stress, as in the crush of humanity during rush hours, are wholly alien to Eskimo experience, yet Eskimos are no strangers to crowding and stress. They experience crowding at the tragic level of starvation in times of scarcity.

6

Spatial Ability, Knowledge, and Place

Animals can move. Agility, speed, and range of motion vary greatly among different species and are largely innate. A newborn ewe, after a few tottering steps, is able to follow its mother about the pasture, managing its four legs so that they do not get in each other's way. Newborn mammals quickly learn to walk. The human young is the well-known exception. A human infant cannot stand or crawl. Even his small bodily movements are rather clumsy. An infant does not quite know where his mouth is, and his first efforts to put his finger in it are a trial-and-error experiment. At a later stage he learns to crawl more or less on his own, but standing and walking—these characteristically human activities—require encouragement and coaching from adults. Spatial ability develops slowly in the human young; spatial knowledge lags further behind. The mind learns to grapple with spatial relations long after the body has mastered them in performance. But the mind, once on its exploratory path, creates large and complex spatial schemata that exceed by far what an individual can encompass through direct experience. With the help of the mind, human spatial ability (though not agility) rises above that of all other species.

Spatial ability becomes spatial knowledge when movements

Spatial Ability, Knowledge, and Place

and changes of location can be envisaged. Walking is a skill, but if I can "see" myself walking and if I can hold that picture in mind so that I can analyze how I move and what path I am following, then I also have knowledge. That knowledge is transferable to another person through explicit instruction in words, with diagrams, and in general by showing how complex motion consists of parts that can be analyzed or imitated.

Since spatial skill lies in performing ordinary daily tasks, spatial knowledge, while it enhances such skill, is not necessary to it. People who are good at finding their way in the city may be poor at giving street directions to the lost, and hopeless in their attempts to draw maps. They have difficulty envisaging their course of action and the spatial characteristics of the environment in which it takes place. There are many occasions on which we perform complex acts without the help of mental or material plans. Human fingers are exceptionally dexterous. A professional typist's fingers fly over the machine; all we see is a blur of movement. Such speed and accuracy suggest that the typist really knows the keyboard in the sense that he can envisage clearly where all the letters are. But he cannot; he has difficulty recalling the positions of the letters that his fingers know so well. Again, riding a bicycle requires muscular coordination and a fine sense of balance, that is to say, a feeling for the distribution of mass and of forces. A physicist may be able to diagram the balance of forces necessary to the mastery of a bicycle, but such knowledge is certainly not required. Self-conscious knowledge can even stand in the way of perfecting a skill.[1]

When we see an animal moving through a long and devious path to reach food or home, it is tempting to ascribe to the animal an experience similar to what our own would be if we were to make the trip. In particular, it is tempting to postulate that the animal envisages a specific goal (a wedge of cheese or a hole in the dining room wall) and that it can picture the path along which it is to travel. This is highly improbable. Even human beings, who are visually and mentally equipped for such acts, rarely find it necessary to exercise their image-making powers.[2]

Spatial Ability, Knowledge,
and Place

We do many things efficiently but unthinkingly out of habit. It is uncanny to watch people acting with skill and apparent purpose and yet know they perform unconsciously, much as our physiological processes adjust to changes in the environment without our conscious control. An extreme example is somnambulism. Perhaps several million Americans sleepwalk or have walked in their sleep when they were children. Lore about somnambulism is abundant. Some of the stories are hard to believe, yet the phenomenon is richly documented and has been studied under laboratory conditions. Here is a striking case. A Berkeley housewife rose one night at 2 A.M., threw a coat over her pajamas, and gathered the family dachshunds into the car for a long drive to Oakland, awakening at the wheel twenty-three miles away.[3] An entire family may be afflicted by somnambulism. When a group of human beings act in consort, with apparent deliberation and purpose, it is hard to accept the fact that they can be quite unaware of what they are doing. Yet several cases of group somnambulism have been reported. One night an entire family of six (husband, his cousin-wife, and their four children) arose at about three in the morning and gathered around the tea table in the servants' hall. One of the children in moving about upset a chair. Only then did they wake up.[4] If a person is afflicted with somnambulism he is likely to start his act as he enters deep sleep. Measurements with electrodes show that sensory information continues to enter the brain of the sleepwalker, whose body then makes appropriate responses, but the brain does not consciously register this information as it does when the person is awake.

With long-distance commuting and driving a commonplace experience in several sectors of American society, many people have known what it is like to display spatial skill and geographical competence in the absence of conscious awareness. The driver "blanks out" while he is on a familiar stretch of the road. He no longer attends to the task; his mind is elsewhere, and yet for long minutes his body maintains control over the vehicle, showing an ability to adapt to minor changes in the environment, as in negotiating broad curves with the

Spatial Ability, Knowledge,
and Place

wheel or pressing down the gas pedal in response to a long upward incline in the road. Griffith Williams has documented several cases of what he calls "highway hypnosis." One driver reported:

I discovered this fact (amnesia) while driving at night from Portland, Oregon to San Francisco, California. The lights of a town approached and I realized that I had been in an almost asleep condition for about 25 miles. Inasmuch as I knew the road I had traveled was not straight, it was apparent that I had negotiated the road, making all the turns, etc. I did not remember the stretch of road at all. I purposely tried it several times after that and found that I could drive miles and miles without memory of it, and while resting. In each case, the moment any driving emergency appeared, I became fully awake.[5]

Acquiring spatial ability, whether it is to ride a bicycle or to find one's way through a maze, does not depend on the possession of a developed cerebral cortex. Pechstein's experiment is convincing. In his experiment rats and human beings are trained to find their way through mazes of identical pattern. Although the training situation is far more familiar to the human than to the rodent subjects, rats learn as quickly and perform almost as well as human beings.[6] Man's large brain is redundant to learning the kinds of skills in path-finding that are essential to the survival of animals.

How do human beings acquire the ability to thread their way through a strange environment, such as unfamiliar city streets? Visual cues are of primary importance, but people are less dependent on imagery and on consciously held mental maps than they perhaps realize. Warner Brown's experimental work suggests that human subjects can learn to negotiate a maze by integrating a succession of tactual kinesthetic patterns. They learn a succession of movements rather than a spatial configuration or map.[7]

Major steps in Brown's experiment are as follows. The subject wears a gadget over his eyes so that he cannot see the maze but can see the upper portions of the larger environment (the room) and also perceive light and noises from outside the room. In the first attempt, the subject stands at the entrance and is aware of its locality. Once he steps into the maze he will

Spatial Ability, Knowledge,
and Place

FROM SPACE TO PLACE: LEARNING A MAZE

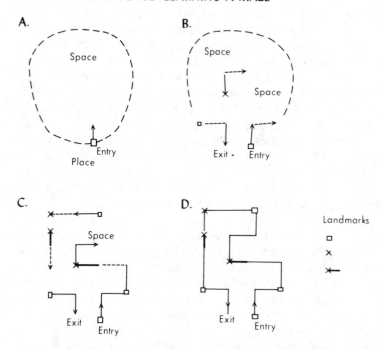

Figure 5. From space to place: learning a maze. At first only the point of entry is clearly recognized; beyond lies space (A). In time more and more landmarks are identified and the subject gains confidence in movement (B, C). Finally space consists of familiar landmarks and paths—in other words, place (D).

move in a certain manner; exit is his goal but its location is not yet known (Fig. 5). By the second or third trial he will have acquired a sense of the location of the exit, and his behavior changes as he approaches it. After a few trials, then, the subject recognizes and expresses confidence about two localities in the laboratory maze, the entrance and the exit. With further trials he learns to identify more and more "landmarks." He refers to them as "a rough spot," "a tilting board," "a long straight stretch," and "double turns." They represent for him stages of a journey. Even past errors can serve this purpose.

Spatial Ability, Knowledge,
and Place

The subject may say, "I made the same mistake last time," indicating that he has recognized a locality. Encountering a familiar landmark in the maze is almost an emotional experience. The subject will often express satisfaction, for the landmark suggests to him that he is on the right track. Moreover each encounter is a signal for what he is to do next. The primary localities remain the entrance and the exit. The integration of space is an incremental process during which the appropriate movements for the entrance and the exit, and for the intermediate localities, continue to expand until they are contiguous. "When the subject is able to tread the maze without error (or with only rare errors) the whole maze becomes *one locality* with appropriate movements."[8] What begins as undifferentiated space ends as a single object-situation or place.

When the person who has learned a maze is asked to walk the same pattern on the open floor, the track he leaves bears only a slight resemblance to the original maze pattern. Elements of the track clearly resemble the correct course, but departures are conspicuous. Most blindfolded subjects, after having learned to tread the maze correctly, fail to apprehend that the plan is rectangular. Few subjects can recount the turns in order as "right" or "left." A subject will attempt to recall the turns and then give up, saying, "I don't know what comes next. I *have to be there* before I can tell you." Drawings of the maze, like the tracks made on the open floor, generally show correct representations of parts of the course, but they are badly executed as to angle and length (Fig. 6). The drawn pattern departs so far from the actual course that it cannot be used as a map.[9]

Brown's experimental work suggests that when people come to know a street grid they know a succession of movements appropriate to recognized landmarks. They do not acquire any precise mental map of the neighborhood. Of course, a rough image of spatial relations can be learned without deliberate effort; people do pick up a sense of the starting point here, the goal out there, and a scattering of intermediate landmarks, but the mental image is sketchy. Precision is not required in the practical business of moving about. A person needs only to have a general sense of direction to the goal, and to know what

Spatial Ability, Knowledge,
and Place

DISTORTION IN DRAWN MAZES

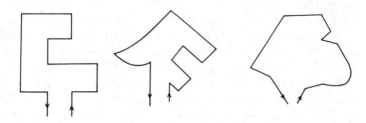

Figure 6. Distortion in drawn mazes. Subjects who have learned how to walk through the maze nonetheless have difficulty reproducing it in drawings. Adapted from Warner Brown, "Spatial integration in a human maze," *University of California Publications in Psychology*, vol. 5, no. 6, 1932, p. 125, 126, figures 2 and 3.

to do next on each segment of the journey. Think what it is like for the man who drives his car from city home to summer cottage. He has made the trip before. At the start he knows the approximate direction of the cottage, so he knows which way to turn the wheel as the car emerges from the driveway. And this knowing what to do next is repeated at each successive landmark; that is, each special configuration in the landscape—which may not always be easy to specify in recapitulation—triggers his next set of movements. D. O. Hebb observes that if the highway curves are gentle the driver has the feeling that, no matter which way he has just swerved his car, he is still heading straight to his destination. A person seems psychologically disposed to discount the angular departures and to accept all forward movements as movements toward his goal.[10] Hence when he tries to reproduce his route in a drawing he is likely to simplify the route and omit or minimize the angularity of the turns—unless he happens to remember a particular bend, in which case he may well exaggerate its angularity.

When space feels thoroughly familiar to us, it has become place. Kinesthetic and perceptual experience as well as the ability to form concepts are required for the change if the space is large. Small children and mentally retarded people are

Spatial Ability, Knowledge,
and Place

likely to have difficulty integrating large space into familiar place. They have no trouble identifying specific landmarks and localities. They recognize particular shops and residences, but they understand the spatial relations among them poorly; hence they easily feel disoriented outside the small areas of habitual contact. We have already considered the world of young children. What problems in orientation confront people who have difficulty forming spatial concepts?

Robert Edgerton gives an example of how confusion can overtake a person (IQ range 55–69) outside the familiar neighborhood. A patient, Mary, is released from a hospital for the mentally retarded so that she can lead as normal a life as possible with the help of a dedicated counselor, Kitty. One day, Mary, Kitty, and Edgerton are on their way to visit Mary's mother. Edgerton reports the following episode:

Kitty was driving and Mary was giving directions. Mary became hopelessly lost. She had no means whatever for finding the house or for explaining her own confusion. The researcher [Edgerton] finally determined the address of the house from a phone book and located the street. They first drove in the wrong direction, as numbers were difficult to see. Mary pointed out a large restaurant as they passed it. The researcher finally noted that they were traveling in the wrong direction, and the car was turned around. Mary quickly pointed out that they must be going in the wrong direction "because there is that restaurant again and we were going wrong when we saw it before." That she was passing the restaurant while traveling in two different directions never occurred to her and she could not be made to understand this fact.[11]

Spatial ability is essential to livelihood, but spatial knowledge at the level of symbolic articulation in words and images is not. Many animals have spatial skills far exceeding those of man; birds that make transcontinental migrations are an outstanding example. For human beings, what is the relationship between spatial ability and knowledge? How does one affect the other? Spatial ability precedes spatial knowledge. Mental worlds are refined out of sensory and kinesthetic experiences. Spatial knowledge enhances spatial ability. This ability is of different kinds, ranging from athletic prowess to such cultural achievements as ocean and space navigation.

Spatial Ability, Knowledge,
and Place

A step in the training of an athlete is to help him envisage the motions necessary for high achievement. Before a man broad-jumps, it helps him to pause and rehearse mentally what he must do. The football coach uses words and diagrams to teach his team the idealized patterns of spatial behavior.[12] Blind people, especially those born without vision, are severely handicapped in movement. To compensate for the lack of vision their auditory and tactile senses are highly developed. Using their minds to formulate spatial concepts further enhances their spatial ability. Tactual maps, for example, help blind children to envisage the relative locations of significant landmarks. From a tactually presented map, boys blind from birth learn to follow a route and even solve a detour problem.[13] Some blind people appear able to use the sun to help them find their way. Verbalizing a course of action is another device that the blind employ when they try to figure out spatial dilemmas.[14]

Human beings are not endowed with an instinctive sense of direction, but under training the ability to stay oriented—even in unfamiliar country—can be acutely developed. De Silva reports the case of a twelve-year-old boy who appears to possess "automatic directional orientation." He does not make deliberate efforts to orient himself and yet is never lost in a strange city. The explanation would seem to lie in early training. His mother at one time gave him orders in cardinal directions rather than in the more usual left and right. She would say, "Get me the brush on the north side of the dresser," or "go and sit in the chair on the east side of the porch." Eventually the boy developed an unusual ability to move in a complicated path for relatively long periods of time and to retain his orientation without paying attention to it.[15]

In a narrow sense, spatial skill is what we can accomplish with our body. Its meaning approximates that of agility. In a broad sense, spatial skill is manifest in our degree of freedom from the tie to place, in the range and speed of our mobility. A trained explorer, equipped with map and compass, can move across a strange country with minimal reliance on his personal experience of the terrain. Technical knowledge has made it possible for human beings to cross continents like birds and indeed, for brief periods, to leave the earth's gravitational field. In people skill and knowledge are inextricably intertwined.

Human groups vary widely in spatial skill and knowledge. Small primitive groups such as the Tasaday of Mindanao occupy a small ecological niche in the forest and hesitate to move beyond their home base. They have no word for sea or lake in their language although these features are less than forty miles away.[16] At the other extreme, the ability to traverse great distances and develop elaborate astronomical and geographical lores are characteristic achievements of people in high civilizations. Indeed such knowledge and capacity are another sign of their overall mastery of the physical environment. It is not, however, a rule that large integrated societies will have greater spatial knowledge than small, loosely organized groups. There are many exceptions. Peasants and subsistence farmers, for example, may have evolved complex social relationships, and they may live in large villages and command ample food supply. Their material culture is more sophisticated than that of primitive hunters and fishers. Yet the spatial skills and knowledge of primitive hunters can far exceed those of sedentary agriculturalists bound to locality. The hunter's skill is not simply one of identifying trails, water holes, and feeding grounds in a broad tract of land, although these capacities are very much in evidence; spatial knowledge extends beyond details of terrain to reference points in heaven, and it may be expressed in the abstract notation of maps.

In Siberia several primitive hunting groups show knowledge of astronomy. The Yakuts, for example, can distinguish with the naked eye stars in the Pleiades not usually seen without a telescope. They show an interest in the number of stars; they say there are many stars in the Pleiades group but only seven large ones. The Buriats, and likewise the natives of northeastern Siberia, make use of the polar star at night and of the sun by day on their hunting trips.[17]

Drawing maps is indubitable evidence of the power to conceptualize spatial relations. It is possible to find one's way by dead reckoning and through long experience with little attempt to picture the overall spatial relations of localities. If an attempt is made to conceptualize, the result may remain mental rather than being transcribed into a material medium. What occasions would call for a map? Perhaps the most common occasion is the need to transmit efficiently geographical knowledge to another person. When someone wants to know how to go to a camp or water hole, the most time-consuming help is to take him there. One can, instead, try to describe the route and the nature of the terrain verbally, but this is always difficult, for language is better suited to the narration of events than to the depiction of simultaneous spatial relations. A sketch map, done quickly on sand, clay, or snow, is by far the simplest and clearest way to show the nature of the country. Cartographic ability presupposes not only a talent for abstraction and symbolization on the part of the primitive cartographer but also a comparable talent in the person who looks on, for he must know how to translate wriggly lines and dots back into real terrain. Sketch maps of this kind will probably depict human habitations and footpaths (to indicate the direction of movement), and such natural features as streams and lakes. They are short-lived. Some, however, have been etched on bark, leather, or wood, and have become part of a people's store of material culture. These maps can be quite elaborate, showing more information than is needed for any particular occasion. They intimate a desire to enshrine communal geographical knowledge in cartographic form. Take the Chukchi maps of the Aradyr delta in northeastern Siberia. They are drawn exceptionally skillfully with reindeer blood on wooden boards. The winding course of the river, the vegetation on the shores, fords, and hunting places are easily seen. "The complicated delta with its numerous islands is faithfully reproduced. Two parallel lines show the shores [of the river]. . . . Many splashes of red on the shores no doubt indicate hills. The map picture is enlivened by hunting and fishing scenes. At one corner is a group of three huts, fishing nets are spread in the middle of the river and a herd of swimming reindeer is shown."[18] The Chukchi maps, in general features, compare

Spatial Ability, Knowledge,
and Place

favorably with a map of the same region made, circa 1900, by the Russian Ministry of the Marine.

Primitive hunters in Siberia learn to conceptualize space and translate their detailed knowledge into the symbolic language of maps. Illiterate Russian peasants, on the other hand, have a poor grasp of spatial relations outside their small world; they have no reason to draw maps, and when asked to do so their efforts are inferior to the works of hunters. What elements in culture, society, and the physical environment affect a people's spatial skills and knowledge? What conditions encourage a people to experience their environment and be aware of it to the degree of trying to capture its essence in words and maps?

John Berry has attempted to find answers by studying two preliterate societies, the Temne of Sierra Leone and the Eskimo of Arctic Canada.[19] The spatial skills of the Eskimo are far superior to those of the Temne. Eskimos possess a large spatial-geometric vocabulary, comparable in richness to that of Western technical man, with which to articulate their world. The Temne, in comparison, have a meager stock of spatial-geometric terms. Eskimos are famed for their soapstone carvings and, more recently, for their work in stencil and block-printing. The Temne produce almost no graphics, sculpture, or decoration. Eskimos are good mechanics. Edmund Carpenter reports that they delight in stripping down and reassembling engines, watches, all machinery. "I have watched them repair instruments which American mechanics, flown into the Arctic for this purpose, have abandoned in despair."[20] The Temne show no special mechanical aptitude. Eskimos are superbly versatile travelers; they use and make maps. Temne farmers lack such skills.

Why the contrast? The physical environments of the two peoples are strikingly different. Temne land is covered with bush and other vegetation, offering a wealth of visual stimuli. Color is vivid: trees and grass vary from light to dark green, and against this green backdrop fruits, berries, and flowers provide flashes of red and yellow. The Eskimo environment is bleak. Moss and lichen in summer give the land a uniform gray-brown cast; snow and ice in winter paint the scene in monotone.

Spatial Ability, Knowledge, and Place

When fog or blizzard appears, land, water, and sky lose all differentiation. It is in this poor and poorly articulated environment that the Eskimos, to survive, have refined their perceptual and spatial skills. When all landmarks disappear in mist and driven snow, Eskimos can nevertheless find their way by observing relationships between the lay of the land, types of snow and ice crack, the quality of the air (fresh or salt), and wind direction. In heavy fog the Arctic navigator establishes his position at sea by the sound of waves beating on land and by checking on the wind.[21] Nature may be hostile and enigmatic, yet man learns to make sense of it—to extract meaning from it—when such is necessary to his survival.

Society has a strong impact on the development of spatial skills. Most Temne are rice farmers and live in villages. Their society is rigid and authoritarian. The men have control over the women, and infractions of marital laws are heavily punished. Discipline of children after weaning is harsh. Formal education is in the hands of the secret societies: the young are taught traditional skills and roles during the months in the bush, after which initiation takes place. Security lies in conforming to the ways of the group, for it is this cohesive unit that confronts nature and extracts from it a livelihood. Since nature is fairly benign, a modest effort suffices to earn a living. The Temne individual is rarely alone. Occasions seldom arise when a farmer faces the task of orienting himself in unfamiliar and inhospitable space. He has no need to make a conscious effort to structure space, since the space he moves in is so much a part of his routine life that it is in fact his "place."

The Temne has his place, knows his place, and is rarely challenged by unstructured space. Far different is Eskimo society in Arctic Canada. Eskimos are hunters and live in family-size hunting camps. Their women enjoy freedom and their children are treated with consistent kindness. Eskimos work alone or with close relatives over a broad territory. Their challenges come from a harsh and fickle environment. The individual does not lean on the power of organized society to overcome nature. He relies on his own ingenuity and fortitude. Compared with Temne farmers, Eskimos are individualistic and venturesome.

Spatial Ability, Knowledge,
and Place

They encourage these traits in their children, who will need
self-reliance to survive. Eskimos have adapted to their inhospitable environment and feel more or less at home in it. That
environment, however, is not yet consistently place. They still
need to cope with unstructured space and they have developed the necessary skills and knowledge to do so success-
fully.[22]

Consider another type of spatial competence—navigation.
Crossing an ocean with the help of magnetic compass and
charts is a highly technical achievement of the European and
Chinese civilizations. The navigational skills and geographical
knowledge of the Pacific Islanders are, in their way, equally
impressive. Geographical knowledge may mean a barely con-
ceptualized familiarity with one's local environment. People
know their own neighborhood well. Geographical knowledge
also means a conscious and theoretical grasp of spatial relations among places that one seldom visits. Pacific Islanders
excel in this more abstract kind of geographical understanding.
An exemplar of such excellence is Captain James Cook's in-
formant, Tupaia, of the Society Islands. He was acquainted
with a world reaching from the Marquesas in the east to Fiji in
the west, a span that is the width of the Atlantic Ocean or
nearly the width of the United States. Tupaia accompanied
Cook to Batavia in the *Endeavour*. At more than "2000 leagues"
from his home and despite the ship's circuitous route, "he was
never at a loss to point to Taheitee," as John Reinhold Forster
(1778) admiringly put it.[23] This expansiveness of geographical
horizon, superbly displayed in Tupaia, is seldom matched by
any other people in the world. It is a fact to counterbalance the
image of primitive peoples as being bound to place, their geographical knowledge deteriorating rapidly into mythology away
from their home grounds.

What conditions favor the exceptional development of spatial skills among the Pacific Islanders? They are similar to those
that foster spatial skills among the Eskimos. To survive, Eskimos must know well large tracts of land and water, since food
obtainable within any small area is insufficient. Pacific Islanders

Spatial Ability, Knowledge, and Place

also need to explore a much larger world than their tiny insular base, but not necessarily because food is inadequate on the island and in the adjoining seas. The reasons for distant travel are subtle enough so that the Islanders themselves are not fully aware of them. The natives of Puluwat, for example, claim that they make trips of 130 and 150 miles to get a special kind of tobacco; yet they need only wait a while for the ships to bring the commodity to them. Visits to distant islands expand the food-supply base, but they also allow people to cement old ties, establish new ones, and exchange ideas. A small community the size of Puluwat cannot achieve its present level of culture without the support of a much larger world. The forging of larger socio-political nets broadens the intellectual horizon, extends the range of choice in goods and in marriage partners, and permits the tiny communities to cope more effectively with natural disasters, notably typhoons.[24]

Puluwatans, like Eskimos, see nature as the arena in which their quintessential virtues and skills are displayed. Eskimos master the arts of hunting in snow and ice fields, Islanders the arts of navigation in unmarked seas. With both groups security and success depend on personal skills and knowledge. Initiative has survival value in a watery world where changes of weather and current have an immediate impact on the small craft. Young Islanders are encouraged to develop curiosity. Like Eskimo children, Puluwatan children are pampered and given much freedom. Boys six or seven years old are taken on long canoe trips. At an impressionable age they imbibe navigational lores and experience the sea and sky in all their moods.[25]

Island navigation combines intimate personal ways of knowing with formal conceptual knowledge. Much that the Islander knows about the sea and navigation is picked up without conscious effort. Not everyone becomes a recognized navigator in Island society, but almost everyone has gone on ocean-crossing trips. He is bound to know how a craft feels as it rides over waves and as it alters course with shifts in current and weather. He learns to detect reefs from the subtle changes in the color of the water, and he learns to read the sky. A recog-

nized navigator's knowledge is more detailed and more consciously acquired than that of the ordinary Islander; nevertheless integral experience rather than deliberative calculation in-
forms the many decisions he has to make in the course of a
long voyage. A navigator needs keen eyes, but he must train
the other senses to a high degree of acuity as well. Sometimes
he will deliberately exclude visual cues in order to concentrate
on other kinds of evidence. This is necessary because the stars
may not be visible, and the wave patterns that provide clues to
direction can be difficult to interpret visually from the level of
the boat. Steering by waves then depends on the motion of the
boat rather than on sighting. A navigator from the Society Islands, Tewake, claimed that "he would sometimes retire to the
hut on his canoe's outrigger platform, where he could lie down
and without distraction more readily direct the helmsman onto
the proper course by analysing the roll and pitch of the vessel
as it corkscrewed over the waves."[26]

Island navigation is also a body of detailed knowledge that
can be taught and learned formally. On Puluwat Atoll instruction begins on land. A senior navigator dispenses a massive
dose of specific information to his students, who sit together in
the canoe house and make little diagrams with pebbles on the
mats that cover the sandy floor. "The pebbles usually represent
stars, but they are also used to illustrate islands and how the
islands 'move' as they pass the canoe on one side or the
other."[27] Here, then, is an example of the use of schematic
diagrams to teach spatial relationships. Learning is not complete until the student "at his instructor's request can start with
any island in the known ocean and rattle off the stars both
going and returning between that island and all the others
which might conceivably be reached directly from there."[28]
What the student eventually acquires is not a long litany of
names but the detailed patterns of stars, islands, and reefs (Fig.
7). "The Puluwatans," says Thomas Gladwin, "pictures himself
and his island in his part of the ocean much as we might locate
ourselves upon a road map." The ocean is a network of sea-
ways linking up numerous islands, not a fearsome expanse of

Spatial Ability, Knowledge, and Place

ETAK: A SYSTEM FOR ORGANIZING SPATIAL DATA

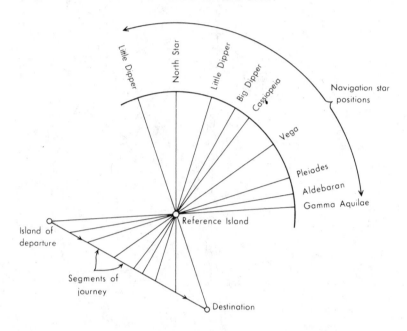

Figure 7. Etak: a system for organizing spatial data. The diagram illustrates the Puluwatan's sophistication in conceptualizing space. "The contribution of etak is not to generate new primary information, but to provide a framework into which the navigator's knowledge of rate, time, geography, and astronomy can be integrated to provide a conveniently expressed and comprehended statement of distance traveled. It also helps keep his attention focused on these key variables which are central to the entire navigation process" (T. Gladwin). Thomas Gladwin, *East is a Big Bird* (Cambridge: Harvard University Press, 1970), opposite page 184. Reprinted with permission from Harvard University Press.

unmarked water (Fig. 8).[29] Polynesian and Micronesian navigators have conquered space by transforming it into a familiar world of routes and places. All people undertake to change amorphous space into articulated geography. Pacific Islanders have reason to take pride in the breadth of their geographical horizon.

Spatial Ability, Knowledge,
and Place

TUPAIA'S IMAGE OF THE PACIFIC OCEAN

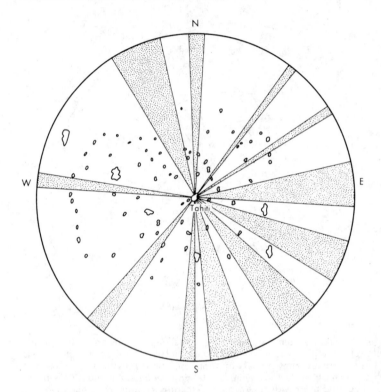

Figure 8. Tupaia's image of the Pacific Ocean. To the Polynesian navigator the ocean is dotted with landmarks. Looking outward from the home base only a small portion of the ocean surface, represented by the shaded sectors, is empty—without islands. Michael Levison, R. Gerard Ward, and John W. Webb, *The Settlement of Polynesia* (Minneapolis: University of Minnesota, 1973), page 63. Figure 37 "Tuapia's map and island screens." Reprinted with permission from the University of Minnesota Press.

7

Mythical Space
and Place

Myth is often contrasted with reality. Myths flourish in the absence of precise knowledge. Thus in the past Western man believed in the existence of the Isles of the Blest, Paradise, the Northwest Passage, and Terra Australis. Now he no longer does. Myths are not, however, a thing of the past, for human understanding remains limited. Political myths today are as common as weeds. Geographical myths, it is true, are less in evidence; we do know much more about the physical characteristics of the earth now than we did before A.D. 1500. But that knowledge is collective: it lies embedded in the great encyclopedias and in works of geography. The knowledge we have as individuals and as members of a particular society remains very limited, selective, and biased by the passions of living.

Myth is not a belief that can be readily verified, or proven false, by the evidence of the senses. The questions posed were not, Is there a Northwest Passage? Is Paradise located in Ethiopia? Rather these places were assumed to exist, and the problem was to find them. Europeans once held tenaciously to the reality of places like the Northwest Passage and a terrestrial paradise. Repeated failure to locate them did not discourage explorers from making further efforts. Such places had to exist because they were key elements in complex systems of belief. To discard the idea of a terrestrial paradise would have threatened a whole way of looking at the world.[1]

Two principal kinds of mythical space may be distinguished. In the one, mythical space is a fuzzy area of defective knowledge surrounding the empirically known; it frames pragmatic space. In the other it is the spatial component of a world view, a conception of localized values within which people carry on their practical activities. Both kinds of space, well described by scholars for nonliterate and traditional societies, persist in the modern world. They persist because for individuals as well as for groups there will always be areas of the hazily known and of the unknown, and because it is likely that some people will always be driven to understand man's place in nature in a holistic way.

Mythical space of the first kind is a conceptual extension of the familiar and workaday spaces given by direct experience. When we wonder what lies on the other side of the mountain range or ocean, our imagination constructs mythical geographies that may bear little or no relationship to reality. Worlds of fantasy have been built on meager knowledge and much yearning. Such stories are often told and need not be repeated here. A less well-known phenomenon is the hazy "mythical" space that surrounds the field of pragmatic activity, to which we do not consciously attend and which is yet necessary to our sense of orientation—of being securely in the world. Think of a man playing with his dog in the study. He sees what lies before him, and through noises and other sensory cues he is aware of the unseen parts of his environment. He is also dimly conscious of what he cannot perceive—for example, the back of the chair when he is not leaning against it and the book-lined wall out of his sight. The man's world does not terminate at the walls of the study; beyond the study lie, successively, the rest of the house, city streets and landmarks, and other cities scattered over the broad face of the nation, all of which are roughly ordered in a compass grid centered on the man himself. Of course the man does not attend to these distant reference points, since he is playing with his dog; yet this large tacit knowledge is necessary to his sense of being at home and oriented in the small arena of activity. When asked, he can envisage the broad field beyond the range of perception; he can make explicit some parts of a large store of tacit knowledge. He may point to the window and say, "Yes, Elm Street is out there and it runs north-and-south." And he may point to a wall and say, "I feel New York is out there, and therefore New Orleans is to my right." Factual errors abound in the unperceived field. This unperceived field is every man's irreducible mythical space, the fuzzy ambience of the known which gives man confidence in the known.[2]

At the level of the cultural group, we have noted that the practical Puluwatans of Micronesia require legendary islands to fill out their space. The Puluwatans are excellent sailors and navigators. Their geographical knowledge extends far beyond their own atoll and the local seas to large parts of the ocean. As a pragmatic people the Puluwatans readily abandon rituals and taboos when these are shown to lack power. Learning how to navigate already puts a heavy burden on memory. Islanders seem glad to drop useless knowledge, yet routes to remote and legendary places are still being taught.[3]

Irving Hallowell, looking for possible human universals, wrote: "Perhaps the most striking feature of man's spatialization of his world is the fact that it never appears to be exclusively limited to the pragmatic level of action and perceptual experience."[4] Hallowell sketched as an example the spatial system of the Saulteaux Indians of Manitoba, who live in the environs of Berens River east of Lake Winnipeg. The terrain they know through direct experience is essentially confined to the winter hunting and summer fishing grounds. Together these make up a small world, but one that the Saulteaux Indian knows in great detail. Beyond this small world, knowledge of terrain becomes hazy and inaccurate. An Indian who works in one hunting area may be ignorant of the geography of another Indian's hunting territory. Yet all Indians will have a rough idea of the locations of the major lakes and rivers far beyond their

Mythical Space and Place

home base, whether they have visited them or not. The small worlds of direct experience are fringed with much broader fields known indirectly through symbolic means.[5]

In contemporary Western society, to give another illustration of a worldwide phenomenon, people in one neighborhood know their own area well but are likely to be ignorant of the area occupied by a neighboring group. Both groups, however, probably share a common store of hazy knowledge (myths) concerning a far larger field—the region or nation—in which their own local areas are embedded. Knowledge of this hazy field is not redundant. Though inaccurate and dyed in phantasms, it is necessary to the sense of reality of one's empirical world. Facts require contexts in order to have meaning, and contexts invariably grow fuzzy and mythical around the edges.

The second kind of mythical space functions as a component in a world view or cosmology. It is better articulated and more consciously held than mythical space of the first kind. World view is a people's more or less systematic attempt to make sense of environment. To be livable, nature and society must show order and display a harmonious relationship. All people require a sense of order and fitness in their environment, but not all seek it in the elaboration of a coherent cosmic system. In general, complex cosmologies are associated with large, stable, and sedentary societies. They are attempts to answer the question of man's place in nature.[6] Practical activities seem arbitrary and may offend the gods or spirits of nature unless they are perceived to have their roles and place in a coherent world system.

How is the human being related to the earth and the cosmos? We shall consider two types of answer, two schemata that are known in widely scattered parts of the world. In one schema the human body is perceived to be an image of the cosmos. In the other man is the center of a cosmic frame oriented to the cardinal points and the vertical axis. We have here two attempts to organize space, not with any narrow purpose in mind, but to gain a sense of security in the universe. The universe is not alien; it influences or determines the fate of human beings and is yet responsive to their needs and initiatives.

The human body is that part of the material universe we know most intimately. It is not only the condition for experiencing the world (chapter 4), but also an accessible object whose properties we can always observe. The human body is a hierarchically organized schema; it is infused with values that are the result of emotion-laden physiological functions and of intimate social experiences. Not surprisingly, man has tried to integrate multifaceted nature in terms of the intuitively known unity of his own body. This perception of an analogy between human anatomy and the physiognomy of the earth is widespread. The Dogon of West Africa see rock as bone, soil as the interior parts of the stomach, red clay as blood, and white pebbles in the river as toes.[7] Certain North American Indian tribes take the earth to be a sentient being made of bones, flesh, and hair. In China popular lore has it that the earth is a cosmic being: mountains are its body, rocks its bones, water the blood that runs through its veins, trees and grass its hair, clouds and mists the vapors of its breath—the cosmic or cloud breath that is life's essence made visible.[8] In the European Middle Ages the idea of the human body as a microcosm was commonplace. As blood vessels permeate the human body so do channels the body of the earth. This view appears in William Caxton's *The Mirrour of the World*, which is translated from a thirteenth-century source.[9] The Elizabethan courtier and adventurer, Sir Walter Raleigh, reiterated the doctrine and added to it the idea that the human breath is analogous to air, and human body heat to the "inclosed warmth" of the earth.[10] This style of reasoning could gain the ear of London's Royal Society as late as the eighteenth century.[11]

The earth is the human body writ large. This makes it easy for traditional thought to comprehend the earth. However, microcosmic theory relates not only the earth but the stars and the planets to the human body. Astrology, according to Barkan and Cassirer, is microcosmic from the beginning. Man is the crucial and central term in the astral cosmos.[12] He contains within him the distillation of the whole astral system. The union of astrol-

Mythical Space and Place

ogy with the body arises out of the need to unite the multiplicity of substances in the universe and out of the search for parallel wholeness. The human body can be inscribed in two cosmic systems: the zodiacal, emphasized by astrologers of the Greco-Egyptian school (like Ptolemy and Manilius), and the planetary, emphasized by the much earlier Chaldeans. In Barkan's words, "Each system has its own kind of power over the individual and its own links to parts of man's body. The zodiacal signs are present at birth and govern the permanent and abiding nature of anatomical features, while the planets, as their configuration changes in the cosmos, determine the day-to-day alterations within our bodies."[13]

Individuals have to work; they must labor in the fields for a living. Agriculture is affected by weather and the seasons, which in turn come under the influence of the stars. In medieval Christian iconography, the zodiacal signs are coupled with seasonal phenomena which alter with the phases of the moon and the sun. Stellar patterns are thus tied to the all-important rhythms of farm work, e.g.:

Aquarius — the coming of rain or the flooding
of a river (January).
Aries — the breeding season of flocks (March).
Virgo — the harvest queen (August).

On church fronts, human beings and their monthly labors are added to the astral signs producing an "anthropodiac," e.g.:

June   Cancer   A mower
July   Leo   A man raking hay
August   Virgo   A harvester[14]

The schema that takes the human body to be an image of the cosmos purports to explain individual human characteristics and fate. For the medieval Christian, astral time and space also affect the produce of the earth and the *timing* of human labor. But the micro-macro cosmic schema does not impose any clear spatial organization on the earth's surface. What is man's place in nature? The answer that has a large spatial component is the second schema, which we shall now consider. The second schema puts man at the center of a world defined by the cardinal points. This idea is perhaps as widespread as the idea of *homo microcosmus*. A spatial frame set in cardinal directions is prominent in New World cosmologies.[15] In the Old World it is well developed in a broad geographical area that stretches from Egypt to India, China and southeast Asia; and beyond these centers of high culture it appears in the simpler economies of interior Asia and of the Siberian plains.[16] From Mediterranean Africa the idea of oriented space may well have penetrated the desert to affect the cosmological thinking of communities in West Africa. The Bushmen in southwest Africa are familiar with the idea of oriented space.[17]

Oriented mythical space differs greatly in detail from one culture to another, but it has certain general characteristics. One is anthropocentrism. It puts man clearly at the center of the universe. The view of the Pueblo Indians of the American Southwest is widely shared. To them, "Earth is the center and principal object of the cosmos. Sun, moon, stars, Milky Way are accessories to the earth. Their function is to make the earth habitable for mankind."[18] Oriented mythical space has other general characteristics. It organizes the forces of nature and society by associating them with significant locations or places within the spatial system. It attempts to make sense of the universe by classifying its components and suggesting that mutual influences exist among them. It imputes personality to space, thus transforming space in effect into place. It is almost infinitely divisible—that is to say, not only the known world but its smallest part, such as a single shelter, is an image of the cosmos.

The individual as well as the shared traits of oriented spatial frames may be illustrated by considering three very different examples: the Saulteaux Indian, the Chinese, and the European. The Saulteaux Indians of Manitoba have a simple hunting and fishing economy. In practical activities they rely exclusively on natural features for orientation, finding their way by means of well-known landmarks. Besides the local landmarks the Indians are aware of the names and the approximate locations of major lakes, rivers, and settlements in North America. This knowledge, we have noted earlier, has no immediate practical value; its use lies in providing a context for what can be perceived. Beyond this contextual field the Saulteaux orient themselves to the North Star, the course of the sun, and the homes of the four winds. These are the reference points of a mythical realm, at the center of which the Indians live. Winds figure prominently in the Saulteaux cosmos. They are anthropomorphic beings, each of whom is identified with a cardinal point. Directions are primarily places—"homes"—rather than courses of movement in space. The idea of space is subordinated to the idea of the location of significant places. The Indians view the east not only as the home of winds, but also as the place where the sun rises; west, the place where it sets. South is the place to which the souls of the dead travel, and the place from which the summer birds come.[19] Cardinal directions are known only approximately. Precise knowledge serves no purpose since the Saulteaux do not use star positions for long distance travel. The construction of a mythical realm satisfies intellectual and psychological needs; it saves appearances and explains events. Do the Indians regard the homes of the winds as having the same substantial reality as, say, Hudson Bay, which they have heard about from travelers? They appear to. Irving Hallowell notes:

There is the Land of the Dead . . . far to the south. There is a road which leads directly to it which deceased souls follow, and a few individuals are known to have visited the Land of the Dead and afterwards returned to their homes. They have given accounts of their journey and of what they saw there. I remember that my interpreter once told an old Indian that I came from the South and that the United States lay in that direction. The old man simply laughed in a wise way and made no comment.[20]

China is an old civilization. In population, economic power, and material attainment the Chinese and the Saulteaux Indians are far apart. Their mythical spaces, however, have points in common. The Chinese, like other peoples, possess a world of empirical knowledge, beyond which extends a circumambient realm of hazy facts and sharply etched legends. Like many North American Indian groups (including the Maya, the Hopi, the Tewa, and the Oglala Sioux), the Chinese put man at the center of space stretched to the four cardinal points, each of which corresponds to a color, and often also to an animal (Fig. 9). Man wants to order his experiences of the world; not surprisingly, the world so ordered revolves around him. The Chinese world view is highly anthropocentric. Consider the design of a roof tile dating back to the Han dynasty. The tile is the Chinese cosmos in miniature. As the earth is rectangular and bounded, so is the tile. Animal symbols lie at the four sides. Close to the east edge is the Blue Dragon, which stands for the color of vegetation and the element wood. Occupying the direction of the rising sun, it is also a symbol of spring. To the south is the Red Phoenix of summer and of fire with the sun at its zenith. To the west is "the White Tiger of the metallic autumn, symbolic of weapons, war, executions, and harvest; of fruitful conclusion and the calmness of twilight, of memory and regret, and unalterable past mistakes."[21] To the north is winter's darkness, out of which all new beginnings must come. North is associated with reptiles, black color, and water. At the center of the cosmos is man on the yellow earth. Man is not pictured on the Han tile, but his very human desires are made known in the written characters for "long life" and "happiness."

In Europe, although a spatial grid of cardinal points existed since antiquity, its role in structuring the cosmos was less important than in certain other centers of high culture such as Central America and China.[22] Whereas the Chinese used a spatial frame of cardinal points to organize the components of nature, the ancient Greeks used the planetary gods. In China, a color, an animal, or an element was attached to each cardinal point. In Greece, a color, a plant, a vowel, a metal, or a stone was attached to each planetary god.[23] A cosmology tied to a spatial frame tended to be more static than one without such a tie. Chinese nature spirits and gods lacked the dynamic and unruly character of Greek gods. Although Occidental culture did not build a well-articulated cosmological system based on the cardinal points, these points nonetheless appeared repeatedly in the conceptualization and construction of various

Mythical Space and Place

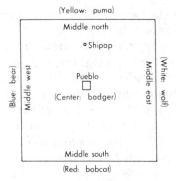

A. A Pueblo Indian world view

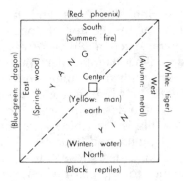

B. Traditional Chinese world view

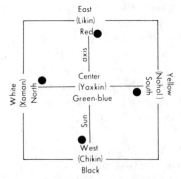

C. Classic Maya world view: quadripartite model of A.D. 600-900

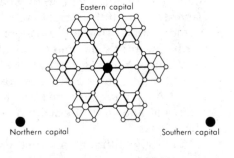

D. Spatial organization of lowland Classic Maya, from regional capital to outlying hamlet: hexagonal model of A.D. 1930

Figure 9. Mythical-conceptual spaces. The world views of the American Indians (A, C) and of the Chinese (B) are alike in that their spatial structure is oriented to the cardinal directions. Spatial organization of the Classic Maya culture reflects its idealized world view (D). Source for C and D: Joyce Marcus, "Territorial organization of the Lowland Classic Maya," *Science*, vol. 180, 1973, figures 2 and 8. Reprinted with permission from Joyce Marcus and the American Association for the Advancement of Science. Copyright 1973 by the American Association for the Advancement of Science.

PTOLEMY'S COSMOS

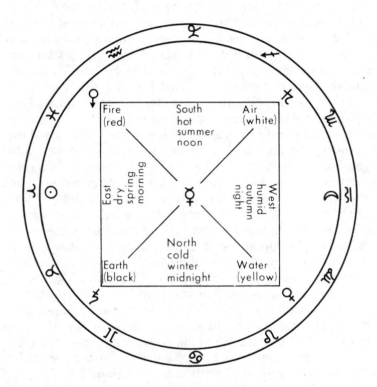

Figure 10. Ptolemy's cosmos. In distinction to the world views of the American Indians and the Chinese (Figure 9), Ptolemy's cosmos subordinates the concept of cardinal points to the heavenly bodies—the zodiacal signs, the sun and the moon, and the planets. Karl A. Nowotny, *Beiträge zur Geschichte des Weltbildes* (Vienna: Ferdinand Berger & Sons, 1970), p. 26. Reprinted with permission from Ferdinand Berger & Sons.

kinds of symbolic space (Fig. 10). In ancient Greece the directions east and west were rich in symbolic meaning. The east connoted light, white, sky, and up; the west suggested darkness, earth, and down. A majority of the post-Doric temples were oriented to the east.[24]

The cosmos of the early medieval period reflects the influence of Greek and pre-Greek thought. In the writings of

Isidore of Seville (c. A.D. 560–636) we find the idea that the universe is divided into four quarters. To Isidore the east quarter was associated with spring, the element air, and the qualities of moisture and heat; west with autumn, earth, dryness, and cold; north with winter, water, cold, and moisture; and south with summer, fire, dryness, and heat.[25] Isidore also recognized four chief winds that came from the four cardinal directions.[26] The four quarters were clearly a major principle for imparting order to space in medieval times.

Christianity had incorporated many of the symbols and rites of pagan antiquity into its own world view. Among the natural symbols for Christ was the sun. The Bible contains such references as "Christ shall shine upon thee" (Ephesians 5:14) and "the sun of righteousness with healing in his wings" shall "arise" on the Day of the Lord (Malachi 4:2). The Christian week begins with Sunday, and in the early years of the church Christians worshiped the resurrected Christ at dawn. In architecture the orientation of the church showed sensitivity to the sun's path since the earliest times.[27] The primacy of the east in the Christian cosmos is evident in the circular (orbis terrarum) maps of the medieval period. On these maps east is shown at the top. Christ's head may be depicted there; his feet appear at the bottom edge in the position of sunset and west; his right hand hovers over Europe and his left hand over Africa. Jerusalem is the navel of Christ and of the world.[28]

We have sketched two spatial schemata: one takes the human body to be a microcosm, the other puts man at the center of a cosmos ordered by the cardinal points. With both schemata an underlying question is, how does the environment affect man, his personality, activities, and institutions? In astrology the influence is sometimes thought to be physical. Heavenly bodies, after all, can be seen to affect tides and weather, so why not also the fate of human beings? A more subtle and mystical relationship is one of "sympathy" between man and the stars. In the Chinese cosmological order, things belonging to the same class affect each other. The process, however, is not one of mechanical causation but rather one of "resonance." For example, the categories east, wood, green, wind, and spring are associated with each other. Change one phenomenon—green, say—and all the others will be affected in a process like a multiple echo. So the emperor has to wear the color green in the spring; if he does not, the seasonal regularity may be upset. The idea here stresses how human behavior can influence nature, but the converse is also believed to occur. Nature affects man: for example, "when the *yin* force in nature is on the ascendancy, the *yin* in man rises also, and passive, negative, and destructive behavior can be expected."[29] Environmental influence is clearly recognized in the cosmological order of the Saulteaux Indians. Thus the winds are not only powers in nature that have to be classified and located in space, they are also active forces in conflict over the middle ground where man lives. North Wind declares that he intends to show no mercy to humans; South Wind, in contrast, intends to treat humans well. The fact that North Wind cannot defeat South Wind in battle means that summer will always return.

Greek astronomers divided the heavens into zones. By the sixth century B.C. the division was also applied to the terrestrial globe. Of the five latitudinal zones transposed to the earth, only the two temperate ones were habitable. Greek geographical science had a strong astronomical component from the beginning. Climate determined; and climate's ancient meaning was "the slope of the sky." By Ptolemy's time the word *klimata* referred to the terrestrial latitude, a measure that could be determined from the elevation of the sun.[30] One heavenly body, the sun, had a pronounced influence on the earth and its inhabitants. People's temperaments and capabilities differed depending on the latitudinal belt in which they lived. Cold regions (north) and warm regions (south) displayed contrasting personalities.

From the fifth century B.C. on, Western thought continued to pursue the themes of environmentalism.[31] Although modern geographers have challenged many of them, some have nonetheless entered folklore and are widely accepted. For example, folk wisdom has it that nations can be divided into "north" and "south": people in the north tend to be hardy and industrious, people in the south tend to be easygoing and artistic. Europe itself is divisible into north and south; each part can claim a distinctive personality. Within Europe, countries such as England, France, Germany, and Italy show marked latitudinal differences. Citizens with any knowledge of their country are seldom at a loss to compare and contrast its two halves in a language that indiscriminately mixes fact with fantasy. Countries have their factual and their mythical geographies. It is not always easy to tell them apart, nor even to say which is more important, because the way people act depends on their comprehension of reality, and that comprehension, since it can never be complete, is necessarily imbued with myths.

The Greeks recognized the sun as the cause of the latitudinal climatic belts on earth. Western thinkers stressed the contrasts between north and south, between cold and hot countries. However, like many other peoples, the Greeks also recognized the sun as the source of life. In the course of time a host of legends accrued to the sun and to its daily passage across the sky. East and west came to be sharply differentiated. East, the place of sunrise, was associated with light and the sky; west, the place of sunset, with darkness and the earth. The right-hand side was identified with the east and the sun, the left-hand side with the "misty west" (*Iliad*). Pythagorean thinkers coupled "right" with "limit," and "left" with the evil of the "unlimited."[32] Isles of the Blest, and later in the Middle Ages the Fortunate Islands, were located in the west. These were idyllic places in which men lived effortlessly, yet such places also connoted death since dead heroes went there. As the sun moved from east to west, so did culture. "Westward the course of empire takes its way," wrote Bishop Berkeley, but he developed an idea that could already be glimpsed in Virgil's *Aeneid*.[33]

Have these traditions of thought survived into modern times? Can one discern the myths of the center and of the cardinal points in the spatial lores of the United States? Cardinal directions, in general, carry little or no symbolic message in the modern world. They can be used simply as a convenient means to differentiate a territory. Australia, for example, is di-

Mythical Space and Place

vided into Western Australia, Northern Territory, and South Australia. The eastern and older part of the island-continent is known under other names. "Western," "Northern," and "South" are labels and no more. Similarly city streets in the United States are qualified by directional terms with no value significance. In Minneapolis an address on "24th Street South" hardly suggests that one lives closer to the sun. But the United States as a whole is divided up into North and South, East and West. Unlike the Australian use of directional terms, regional labels in the United States are not promulgated by central authority; like the regions of mythical space the names and meanings of American regions are acquired in the course of time, as part of the growing lore and literature of a people.[34] In the mythical space of traditional societies, cardinal points are tied to astronomical events and to the seasons with their control over life and death. American space is not a stage set for the enactment of cosmic drama, but, as regional novels and literature show, the physical environment, particularly climate, does play a large role in giving personality to such regions as the South, the Northeast, and the West.[35] In the mythical space of traditional societies the idea of center or "middle place" is important. The idea of a center or heartland is also important to American space.[36] But movement is another key theme in American history. The movement of the people to the west, combined with the powerful lure of the West as an ideal, distorts the sense of symmetry that the concept of center imparts. Hence the term "Middle States" is short-lived. Heartland America is not known as the Middle States but as the Middle West.

Mythical space is an intellectual construct. It can be very elaborate. Mythical space is also a response of feeling and imagination to fundamental human needs. It differs from pragmatic and scientifically conceived spaces in that it ignores the logic of exclusion and contradiction. Logically a cosmos can have only one center; in mythical thought it can have many centers, although one center may dominate all the others.[37] Logically the whole is made up of parts, each with its characteristic location, structure, and function. The part may be essential to the functioning of the whole, but the part is not the whole in miniature and in essence. In mythical thought the part can symbolize the whole and have its full potency. The limits of the Puebloan cosmos are the distant mountains, but they are also the walls of the kiva and of the individual houses. In China, as we have seen, a single roof tile encapsulates the essential order and meaning of the Chinese cosmos. The mythical space depicted there recurs in the house, of which the tile is a part; in the city, of which the house is a part; and finally in the empire, of which the city is a part.

The small mirrors the large. The small is accessible to all human senses. Its messages, being confined within a small area, are readily perceived and understood. Architectural space—a house, a temple, or a city—is a microcosm possessing a lucidity that natural features lack. Architecture continues the line of human effort to heighten awareness by creating a tangible world that articulates experiences, those deeply felt as well as those that can be verbalized, individual as well as collective.

8

Architectural Space
and Awareness

Many animals, like human beings, live in environments of their own construction rather than simply in nature. And evolutionarily advanced animals such as birds and mammals are not the only species that can build. Even single-celled organisms construct shells for themselves out of things like sand grains. We say, however, that animals build instinctively, that each species of weaverbird has an inherited instinct to make a nest of a particular shape, some round, others pear-shaped. Yet we know that some weaverbirds build better nests in their second year than they did in the first. Weaverbirds are capable of learning from experience, which means that not all the details of their performance are controlled by heredity. As another illustration of architectural prowess, consider the termites. They live in a built environment that is vast in proportion to their own size. They make nests that soar like skyscrapers. Termites' nests contain not only elaborate ventilated living quarters for themselves but also fungus gardens for their form of food production. Moreover there appear to be local traditions in architecture that determine how, for instance, the ventilation should be arranged; termites of the same species adopt different systems in Uganda and on the west coast of Africa.[1]

Compared with the termite's skyscraper, the lean-tos and

Architectural Space
and Awareness

thatched mud shelters of the human being look crude. If humans nonetheless claim a certain superiority, the claim must rest on grounds other than architectural achievement. It must rest on awareness. The assumption is that the Bushman, when he makes his lean-to shelter, is more aware of what he does than the weaverbird and the termite as they make their fancier homes.

What is the quality of this awareness? What is the human builder conscious of as he first creates a space and then lives in it? The answer is complex because several kinds of experience and awareness are involved. At the start, the builder needs to know where to build, with what materials, and in what form. Next comes physical effort. Muscles and the senses of sight and touch are activated in the process of raising structures against the pull of gravity. A worker modifies his own body as well as external nature when he creates a world. Completed, the building or architectural complex now stands as an environment capable of affecting the people who live in it. Man-made space can refine human feeling and perception. It is true that even without architectural form, people are able to sense the difference between interior and exterior, closed and open, darkness and light, private and public. But this kind of knowing is inchoate. Architectural space—even a simple hut surrounded by cleared ground—can define such sensations and render them vivid. Another influence is this: the built environment clarifies social roles and relations. People know better who they are and how they ought to behave when the arena is humanly designed rather than nature's raw stage. Finally, architecture "teaches." A planned city, a monument, or even a simple dwelling can be a symbol of the cosmos. In the absence of books and formal instruction, architecture is a key to comprehending reality. Let us look at these kinds of experience and awareness in greater detail.

Where shall one build, with what materials, and in what form? Such questions, it has been said, do not worry builders in preliterate and traditional societies. They work from ingrained habit, following the procedure of unchanging tradition. They have, in any case, little choice since both the skill and the materials at hand are limited. Some types of dwelling, such as the beehive houses of Apulia, the black houses of the Outer Hebrides, and the Navajo hogans, have not changed since prehistoric times. Habit dulls the mind so that a man builds with little more awareness of choice than does an animal that constructs instinctively. At the opposite pole from the primitive builder is the modern master architect. He feels the call to be original. He can, if he likes, select and combine from the numerous styles offered by the world's cultures, past and present. He has almost unlimited technical means at his disposal to achieve his final vision. Given a project, the master architect is obligated to conceive in his mind and on paper a range of architectural forms, all of which serve the project's purpose but only one of which will be selected because it is deemed the best, for reasons that may not be clear to the architect himself.[2] In the preliminary steps of design the architect's consciousness is almost painfully stretched to accommodate all the possible forms that occur to him.

This contrast between primitive builder and modern architect is, of course, an exaggeration: the one is not wholly chained to custom and the other does not have unlimited choice. What sorts of decisions does the primitive builder make? What are his options? These are pertinent questions because a person is most aware when he has to pause and decide. Unfortunately we lack the evidence for clear answers. Few ethnographic surveys report on building activity as a process of making up minds, of communication and learning. Rather huts and villages are described as though they simply appeared, like natural growths, without the aid of cogitating mind. Such portraitures are, to say the least, misleading. In any human life choices arise and decisions must be made, even if they are not especially demanding. Nomads, for example, need to decide where to stop for the night, where to establish their camps. Shifting agriculturalists must know where to make a clearing and build a village. These are locational choices. Material and form also require selection. The natural environ-

Architectural Space
and Awareness

ment is never static or uniform. Materials available to the human builder vary, however slightly, in time and place, forcing him to think, adjust, innovate.

Nonliterate and peasant societies are conservative. Their shelters show little change in the course of time, and yet— paradoxically—there may be greater awareness of built forms and space in a traditional than in a modern community. One cause of such greater awareness is active participation. Since nonliterate and peasant societies do not have architects, everyone makes his own house and helps to build public places. Another cause is that this effort, with the awareness it stimulates, is likely to be repeated many times in the course of a man's life. Primitive shelters combine persistence of form with ephemerality of substance. Construction and repair are almost a constant activity. A house is not achieved once to be enjoyed thereafter. The Eskimo, on winter hunts, makes a new igloo every night. The Indian's tepee rarely lasts more than one season. Every few years the shifting cultivators must clear another patch of forest and build another village.

A third cause of heightened awareness is the fact that with many primitive and traditional peoples the act of construction is a serious business that calls for ceremonial rites and perhaps sacrifice.[3] To build is a religious act, the establishment of a world in the midst of primeval disorder. Religion, since it is concerned with stable truths, contributes to the conservatism of architectural form. The same shaped house and city are made again and again as though they come out of the mold of some unthinking process of mass production; yet each is probably built with a sense of solemnity. The builder, far from feeling that he is doing routine work, is obliged by the ceremony to see himself as participating in a momentous and primordial act. The occasion elevates feeling and sharpens awareness, even though the actual steps to be followed in construction fall into a more or less prescribed pattern.

A type of spatial consciousness that people of a simple economy do not experience is systematic and formal design, the envisagement of the final result by drawing up plans. Any large enterprise calls for conscious organization. This can be done verbally and by example on the work site. However, an order of complexity develops, at which point instruction has to be more formally presented if it is to be effective. A technique in formal learning and teaching is the plan or diagram. By making sketches the architect clarifies his own ideas and eventually arrives at a detailed plan. With the same means he helps others to understand what is to be done. The plan is necessary to any architectural enterprise that is sustained over a period of time and executed by a large team of more or less specialized workers. Conceptualizing architectural space with the help of plans is not, of course, a modern device. According to John Harvey, from Egypt in the middle of the second millennium B.C. there is a continuous chain of evidence for architectural scale drawing, throughout all the higher cultures of the Near and Middle East and in classical and medieval Europe.[4]

Indeed by late medieval times the prototype of the modern architect appeared in Europe. He was the master builder, a man of vision and temperament who did not hesitate to impose his own personality on design. The master builder had a certain freedom of choice; he could, for example, select the fashionable Gothic arch as against the outmoded Romanesque. Size was another area that allowed a certain latitude. A building might serve a traditional purpose and yet permit the architect to exercise initiative, for to construct a monument at all revealed hubris, that is, a yearning to excel, to depart from precedence if only in size and in decorative conceits. Wealthy patrons might share the megalomania of their architects. Instructions tended to be general rather than specific. Abbot Gaucelin (A.D. 1005–1029) at Fleury decided to build a tower at the west end of the minster with the square stones he had brought by boat from the Nivernais. His simple instruction to his architect was, "Build it to be a model for the whole of France."[5]

The late medieval period had known cultural innovation, conspicuously so in monumental architecture. At the same time Christian values remained intact and formed a bond for people in different walks of life. The construction of a cathedral aroused the enthusiasm of a broad community of believers. When Chartres was being built, Robert of Torigni reported glowingly that 1,145 men and women, noble and common people, together dedicated all their physical resources and spiritual strength to the task of transporting in hand-drawn carts material for the building of the towers.[6] Such accounts suggest that raising an edifice was an act of worship in which the feelings and senses of a people were deeply engaged. The vertical structure of the medieval cosmos was not then an abstract and dry doctrine that had to be accepted on faith but rather a world that could be seen and felt as the arches and towers heaved heavenward. In the sixteenth century an architectural enterprise dedicated to God could still inspire a fervor among workers and populace that we in our secular age may find incomprehensible. Here is the historian Leopold von Ranke's description of the raising of the obelisk in front of St. Peter's on April 30, 1586.

[Raising the obelisk] was a work of the utmost difficulty. All who were engaged in it seemed inspired with the feeling that they were undertaking a work which would be renowned through all the ages. The workmen, nine hundred in number, began by hearing Mass, confessing, and receiving the Communion. They then entered the space which had been marked out for the scene of their labors. The master placed himself on an elevated seat. The obelisk was bound with strong iron hoops. Thirty-five windlasses were to set in motion the monstrous machine which was to raise it with strong ropes. At length a trumpet gave the signal. At the very first turn, the obelisk was heaved from the base on which it had rested for fifteen hundred years. At the twelfth, it was raised two palms and a quarter, and remained steady. The master saw the huge mass in his power. A signal was fired from Fort St. Angelo, all the bells in the city rang, and the workmen carried their master in triumph around the inclosure, with incessant shouts and acclamations.[6]

Building is a complex activity. It makes people aware and take heed at different levels: at the level of having to make pragmatic decisions; of envisioning architectural spaces in the mind and on paper; and of committing one's whole being, mind and body, to the creation of a material form that captures an ideal. Once achieved, architectural form is an environment for man. How does it then influence human feeling and consciousness? The analogy of language throws light on the question. Words contain and intensify feeling. Without words feeling reaches a momentary peak and quickly dissipates. Perhaps one reason why animal emotions do not reach the intensity and duration of human ones is that animals have no language to hold emotions so that they can either grow or fester. The built environment, like language, has the power to define and refine sensibility. It can sharpen and enlarge consciousness. Without architecture feelings about space must remain diffuse and fleeting.

Consider the sense of an "inside" and an "outside," of intimacy and exposure, of private life and public space. People everywhere recognize these distinctions, but the awareness may be quite vague. Constructed form has the power to heighten the awareness and accentuate, as it were, the difference in emotional temperature between "inside" and "outside." In Neolithic times the basic shelter was a round semisubterranean hut, a womblike enclosure that contrasted vividly with the space beyond. Later the hut emerged above ground, moving away from the earth matrix but retaining and even accentuating the contrast between interior and outside by the aggressive rectilinearity of its walls. At a still later stage, corresponding to the beginning of urban life, the rectangular courtyard domicile appeared. It is noteworthy that these steps in the evolution of the house were followed in all the areas where Neolithic culture made the transition to urban life.[8]

The courtyard house is, of course, still with us—it has not become obsolete. Its basic feature is that the rooms open out to the privacy of interior space and present their blank backs to the outside world. Within and without are clearly defined; people can be certain of where they are. Inside the enclosure, undisturbed by distractions from the outside, human relations and feelings can rise to a high and even uncomfortable level of warmth. The notion of inside and outside is familiar to all, but imagine how sensibly real these categories become when a guest—after a convivial party—leaves the lantern-lit courtyard and steps through the moon gate to the dark wind-swept lane outside. Experiences of this kind were commonplace in tradi-

Architectural Space
and Awareness

tional Chinese society, but they are surely known to all people who use architectural means to demarcate and intensify forms of social life (Fig. 11). Even contemporary America, with its ideal of openness symbolized by large windows and glass walls, has created the enclosed suburban shopping center. How will the shopper experience such a place? As he approaches it in his car across the vast expanse of the parking lot, he can see only the center's unperforated outer sheath which, except for a large trade sign, makes no attempt to lure people in. The image is bleak. He parks the car, steps inside the center's portal, and at once enters a charmed world of light and color, potted plants, bubbling fountains, soft music, and leisurely shoppers.[9]

Spatial dimensions such as vertical and horizontal, mass and volume are experiences known intimately to the body; they are also felt whenever one sticks a pole in the ground, builds a hut, smoothes a surface for threshing grain, or watches a mound of dirt pile up as one digs a deep well. But the meaning of these spatial dimensions gains immeasurably in power and clarity when they can be seen in monumental architecture and when people live in its shadow. Ancient Egypt and Mesopotamia have enlarged mankind's consciousness of space, heightened people's awareness of the vertical and the horizontal, of mass and volume, by constructing their exemplars in the towering shapes of pyramids, ziggurats, and temples.[10] We have inherited this knowledge. Modern architects design with these dimensions in mind. The layman, sensitized to the dramatic play of thrust and repose, learns to appreciate it wherever it appears, in nature as well as in man-made objects that have no aesthetic pretension. We see drama and meaning in the volcanic neck standing above the flat plateau and in the silos of Nebraska. Here is Wright Morris's description of the symbols in a prairie town. The grain elevators are, for him, the monuments of the plains. He observes:

There's a simple reason for the grain elevators, as there is for everything, but the forces behind the reason, the reason for the reason, is the land and the sky. There is too much sky out there, for one thing, too much horizontal, too many lines without stops, so that the excla-

INTERIOR SPACE AND THE COURTYARD HOUSE

A. House at Ur ca. 2000 B.C.

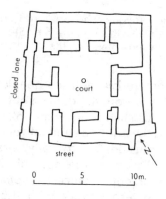

B. House at Oaxaca, Mexico ca. A.D. 600

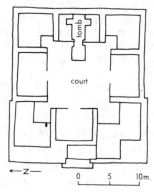

C. House at Priene ca. 300 B.C.

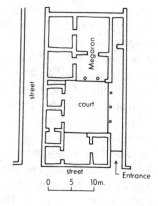

D. A typical Peking house

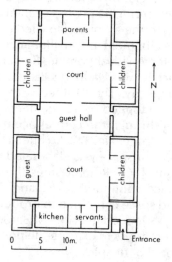

Figure 11. Interior space and the courtyard house. Here is a type of domestic environment that transcends time and culture. The courtyard house dramatizes the contrast between "inside" and "outside." Source for B: Marcus C. Winter, "Residential patterns at Monte Alban, Oaxaca, Mexico," *Science*, vol. 186, 1974, page 985, figure 5. Reprinted with permission from Marcus C. Winter and the American Association for the Advancement of Science. Copyright 1974 by the American Association for the Advancement of Science. Figure 11D is adapted from Andrew Boyd, *Chinese Architecture and Town Planning* (Chicago: University of Chicago Press, 1962), page 80, figure 29.

Architectural Space
and Awareness

mation, the perpendicular had come. Anyone who was born and raised on the plains knows that the high false front on the Feed Store, and the white water tower, are not a question of vanity. It's a problem of being. Of knowing you are there.[11]

A third example of how architecture can educate people's awareness and conception of reality is from the domain of the illuminated interior. Interior space as such is a commonplace experience. We have already noted the enduring and universal antithesis between "inside" and "outside." Historically, interior space was dark and narrow. This was true not only of humble dwellings but also of monumental edifices. Egyptian and Greek temples commanded external space with their polish and imposing proportions; their interiors, however, were gloomy, cluttered, and crudely finished. European architectural history has seen many changes of style but, according to the art historian Giedion, among ambitious builders the development of an illuminated and spacious interior was a common ideal from the Roman to the Baroque period. An early success was Hadrian's Pantheon. Its interior attained a sublime simplicity. The Pantheon consisted essentially of a cylindrical drum topped by a large hemispherical dome; sunlight streaming through the central oculus swept the building's stark hollow space (Fig. 12). Architectural drawings and relics show that interior space was elaborated together with the fenestration of light. From Roman times the role of light in defining interior space continued to expand. With the Gothic cathedral light and space combined to produce effects of mystical beauty. The light-flooded interiors of Baroque churches and halls were further efforts to explore the possibilities of a major and enduring concept of space.[12]

In sketches of architectural development like these, we trace the growth of the human capacity to feel, see, and think. Woolly feelings and ideas are clarified in the presence of objective images. Perhaps people do not fully apprehend the meaning of "calm" unless they have seen the proportion of a Greek temple against the blue sky, or of "robust, vital energy" without baroque façades, or even of vastness without a huge edifice.[13] But, we may well ask, doesn't nature provide even

*Architectural Space
and Awareness*

THE DOME: EXPERIENCE AND SYMBOL

A. Mongolian yurt

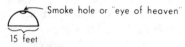

Smoke hole or "eye of heaven"

15 feet

B. Hadrian's Pantheon

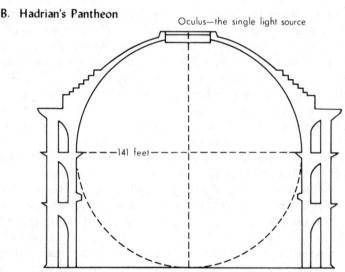

Oculus—the single light source

141 feet

Figure 12. The dome: experience and symbol. The symbolism of the modest Mongolian yurt resembles that of one of the world's architectural masterpieces—Hadrian's Pantheon: the dome is the vault of heaven and the central opening is the "eye of heaven." Yet how vastly different is the experience: entering the smoke-stained yurt can hardly evoke the same feeling as entering the vaulting interior of the Pantheon. Adapted from Sigfried Giedion, *Architecture and the Phenomena of Transition* (Cambridge: Harvard University Press, 1971), page 148, figure 116.

more powerful images? What gives a better sense of calm than the sea at rest, or of exuberant energy than the primeval forest, or of vastness than the endless sweep of the plains? True, but it is doubtful whether human beings can naïvely apprehend these qualities in nature without prior experience in the sensible forms and scale created by man. Nature is too diffuse, its stimuli too powerful and conflicting, to be directly accessible to the human mind and sensibility. First man creates the circle, whether this be the plan of the tepee or the ring of the war dance, and then he can discern circles and cyclical processes everywhere in nature, in the shape of the bird's nest, the whirl of the wind, and the movement of the stars.[14]

The designed environment serves an educational purpose. In some societies the building is the primary text for handing down a tradition, for presenting a view of reality. To a nonliterate people the house may be not only a shelter but also a ritual place and the locus of economic activity. Such a house can communicate ideas even more effectively than can ritual. Its symbols form a system and are vividly real to the family members as they pass through the different stages of life (Fig. 13).[15] On a larger scale the settlement itself may be a potent symbol. Consider villages on the island of Nias in Indonesia. A South Nias village is a diagram of cosmic and social order. Its characteristic location is the hilltop. The word for village also means "sky" or "world." The chieftain is called "that which is up river." His large house, located at the upper end of the central street, dominates the settlement. The street's upper end signifies river source, east or south, the sun, aerial creatures, chieftainship, and life. The lower end signifies downstream, west or north, aquatic animals, commoners, and death. A man's status is clearly indicated by the size and location of his house. Slaves live either in the field, beyond the cosmic village, or under the village dwellings and share space with pigs. Such a world would constantly remind man of where he stands both in society and in the cosmic scheme of things.[16]

In nonliterate and traditional communities the social, economic, and religious forms of life are often well integrated. Space and location that rank high socially are also likely to have religious significance. An economic activity may be deemed profane, but "profane" is itself a religious concept. In contrast, modern life tends to be compartmentalized. Space in our contemporary world may be designed and ordered so as to draw one's attention to the social hierarchy, but the order has no religious significance and may not even correspond closely to wealth. One effect is the dilution of spatial meaning. In mod-

*Architectural Space
and Awareness*

COSMIC AND SOCIAL ORDER IN ATONI HOUSE

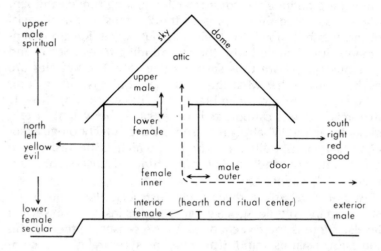

Figure 13. Cosmic and social order in the Atoni house of Indonesian Timor. "Order in building expresses ideas symbolically, and the house depicts them vividly for every individual from birth to death. Furthermore, order concerns not just discrete ideas or symbols, but a system; and the system expresses both principles of classification and a value for classification per se, the definition of unity and difference" (C. Cunningham). Clark E. Cunningham, "Order in the Atoni House," in Rodney Needham, ed., *Right and Left: Essays in Dual Symbolic Classification* (Chicago: The University of Chicago Press, 1973), page 219, figure 7. Reprinted with permission from the University of Chicago Press. Copyright 1973 by the University of Chicago.

ern society spatial organization is not able, nor was it ever intended, to exemplify a total world view.

All settlement patterns reveal at least social order and functional convenience. Other types of order may or may not be superposed. Space of restricted meaning is characteristic of, but not limited to, Western technological society. It appears also among people with the simplest economy and social structure. An example is provided by the Pygmies of the Congo forest. Their created space is the camp, a forest clearing with conical huts arranged in a scalloped ring. Friends build their huts so that the openings face each other. A man speaks as an individual when he talks from his own doorway, but he speaks for the group when he stands in mid-camp.[17] The center is public, the periphery is for interaction among friends and kin. Man-made space thus expresses the Pygmies' informal social order. It is not, however, their religious space. Religious sentiments are identified with the surrounding forest. Social and religious spaces are thus separate (Fig. 14). The Pygmies are perhaps aware that what they make and build is trifling compared with the circumambient and life-supporting forest: man-made things cannot, as it were, carry the weight of religious meaning. What distinguishes Western technological society is that its built environment, which is pervasive and dominant, nonetheless has only minimal cosmic or transcendental significance.

Architectural space reveals and instructs. How does it instruct? In the Middle Ages a great cathedral instructs on several levels. There is the direct appeal to the senses, to feeling and the subconscious mind. The building's centrality and commanding presence are immediately registered. Here is mass—the weight of stone and of authority—and yet the towers soar. These are not self-conscious and retrospective interpretations; they are the response of the body. Inside the cathedral there is the level of explicit teaching.[18] Pictures in the stained-glass windows are texts expounding the lessons of the Bible to illiterate worshipers. There are the countless signs pointing to Christian doctrine, practice, and mystery: holy water, flickering candle light, statues of saints, confessional, pulpit, altar, and cross are examples. To some of the signs the worshipers respond with a more or less automatic act, such as kneeling. Other signs elicit specific ideas. The cross suggests suffering, atonement, and salvation. Finally the cathedral as a whole and in its details is a symbol of paradise. The symbol, to the medieval mind, is more than a code for feelings and ideas that can readily be put into words. The symbol is direct and does not require linguistic mediation. An object becomes a symbol when its own nature is so clear and so profoundly exposed that while being fully itself it gives knowledge of something greater beyond. Imagine a man of the Middle Ages who goes into a cathedral to worship and meditate. He is reverent and has

PYGMY CAMP: SOCIAL AND SACRED SPACE

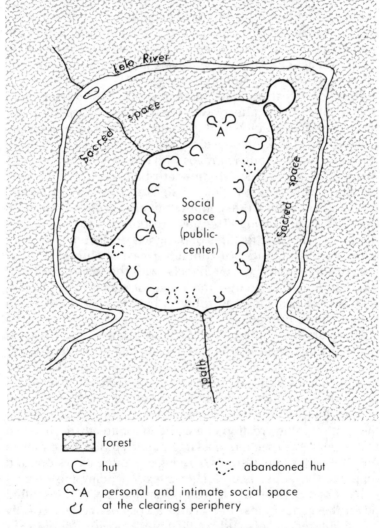

forest

⊂ hut   ⊙ abandoned hut

⊊A   personal and intimate social space
⊍   at the clearing's periphery

Figure 14. Pygmy camp in the Ituri (Congo) rainforest, showing personal, social, and sacred space. Adapted from Colin M. Turnbull, "The lesson of the Pygmies," *Scientific American*, vol. 208, 1963, p. 8.

some learning; he knows about God and heaven. Heaven is that which towers over him, has great splendor, and is suffused in divine light. These are, however, only words. In an ordinary setting, when he tries to envisage paradise by the power of his own imagination his success is likely to be modest. But in the cathedral his imagination need not soar unaided. The beauty of space and light that he can perceive enables him to apprehend effortlessly another and far greater glory.[19]

Turn now to earth and the modern world. How does modern architectural space affect awareness? In important respects, the principal ways by which it influences people and society have not changed. Architectural space continues to articulate the social order, though perhaps with less blatancy and rigidity than it did in the past. The modern built environment even maintains a teaching function: its signs and posters inform and expostulate. Architecture continues to exert a direct impact on the senses and feeling. The body responds, as it has always done, to such basic features of design as enclosure and exposure, verticality and horizontality, mass, volume, interior spaciousness, and light. Architects, with the help of technology, continue to enlarge the range of human spatial consciousness by creating new forms or by remaking old ones at a scale hitherto untried.

These are the continuities. What are some of the changes? Active participation is much reduced. In the modern world people do not, as in nonliterate and peasant societies, build their own houses, nor do they participate even in a token manner in the construction of public monuments. Rites and ceremonies that focus on the building activity, which used to be thought of as the creation of a world, have greatly declined so that even in the erection of a large public edifice there remain only the rather wan gestures of laying the foundation stone and topping. The house is no longer a text encoding the rules of behavior and even a whole world view that can be transmitted down the generations. In place of a cosmos modern society has splintered beliefs and conflicting ideologies. Modern society is also increasingly literate, which means that it depends less and less on material objects and the physical environment to embody the value and meaning of a culture: verbal symbols have progressively displaced material symbols, and books rather than buildings instruct.

Symbols themselves have lost much of their power to reverberate in the mind and feeling since this power depends on the existence of a coherent world. Without such a world symbols tend to become indistinguishable from signs. Gas stations, motels, and eateries along the highway have their special signs which are intended to suggest that these are not only convenient but good places for the motorists to pause. Holiday Inn's trademark promises room, food, and service of a certain quality.[20] What else does it say? We can of course think of other values, but a characteristic of the live symbol is that it does not require explication. Consider the modern skyscraper. People who take note of it are likely to offer a broad range of opinions concerning its worth and meaning. To some it is aggressive, arrogant, and monolithic; to others, on the contrary, it is daring, elegant, and lithe. Such divergent—even opposing—views exist despite the fact that the high-rise is the product of an age to which we all belong. A *consensus gentium* is notably lacking with regard to the artifacts of modern culture. Turn again to the Gothic cathedral. As with the modern skyscraper it is capable of provoking divergent opinions. It has been called "an expression of ignorant and monkish barbarians," "the finest utterance of a noble faith," "the architectural image of primeval forests," and "the lucid embodiment of constructive mathematics."[21] But what is sampled here are the literary views of critics who lived in later times. To those who built the cathedral and to the faithful who worshiped in it, the edifice probably did not require further literary exposition. In that age of concrete symbols people could accept it as the forecourt to paradise, an artifact handsome in itself and yet revelatory of a far more exalted realm.

9

Time
in Experiential
Space

Our discussion of space and place has thus far made no explicit reference to time, which is, however, implied everywhere in the ideas of movement, effort, freedom, goal, and accessibility. The purpose of this chapter is to relate time explicitly to space.

The experience of space and time is largely subconscious. We have a sense of space because we can move and of time because, as biological beings, we undergo recurrent phases of tension and ease. The movement that gives us a sense of space is itself the resolution of tension. When we stretch our limbs we experience space and time simultaneously—space as the sphere of freedom from physical constraint and time as duration in which tension is followed by ease. The readiness with which we confound spatial and temporal categories is apparent in the language. Length is commonly given in time units. Architectural space, because it can seem to mirror rhythms of human feeling, has been called "frozen music"—spatialized time. The passage of time, conversely, is described as "length." Time is even "volume," as, for example, when people speak of having a "big time," a figure of speech that, according to Langer, is psychologically more accurate than to talk of having a busy or exciting time.[1] Daily living in modern society requires that we be aware of space and time as separate dimensions and as transposable measures of the same experience. We wonder whether there is parking space, whether we shall be late for an appointment, and even as we estimate the distance from parking lot to office in terms of time we wish we had been able to assign a bigger block of time for the appointment.[2]

People differ in their awareness of space and time and in the way they elaborate a spatio-temporal world. If people lack a sense of clearly articulated space, will they have a sense of clearly articulated time? Space exists in the present; how does it acquire a temporal dimension? Consider the possibility that the environment itself may have an effect on the elaboration of a spatio-temporal world. Natural environments vary conspicuously over the earth's surface and cultural groups differ in the way they perceive and order their environments; yet almost everywhere people distinguish two types of space, the land and the sky. The Congo Pygmies are a striking exception.[3] Because they are enveloped in the dense forest, the distinction "land" and "sky" lacks perceptual support for them. The sky is seldom visible. The sun, the moon, and the stars, from which many societies derive their measure of recurrent time, can rarely be seen. Vegetation camouflages all landmarks. A Pygmy cannot stand on a prominence and survey space before him; he cannot peer into the horizon where events occurring then may affect him later. He does not learn to translate spontaneously the apparent size of an object into distance. For example, he tends to see a distant buffalo as a very small animal.[4] Distance, unlike length, is not a pure spatial concept; it implies time.

In a dense forest environment, what can distance mean? Aural cues give a sense of distance, but sounds present a smaller world than what eyes can potentially see. Moreover, whereas visual space tends to be focused and structured around an object or a succession of objects, aural space is less focused. Forest sounds are not precisely located; they yield an ambience rather than a coordinated spatial system. Space, to the rain forest dwellers, is a dense net of places with no overall structure. The same is apparently true of time. The negligible seasonal rhythm deprives the forest people of a measure and a concept of time overarching the quick successions of the diurnal period. The time span known to the Pygmies is restricted. Although they have a detailed knowledge of many plants and animals, they pay little heed to life as stages of growth. Time, like perceived distance, is shallow: neither the genealogical past nor the future holds much interest.

The Hopi Indians of the American Southwest live on a semi-arid plateau. In the clear and dry air they can see into great distances. Their environment is one of panoramic views and sharply etched landmarks in antithesis to the womblike, sheltered milieu of the rain forest dwellers. How are space and time perceived and integrated in the Hopi world? According to Benjamin Whorf, the Hopi recognize two realms of reality: manifested (objective) and manifesting (subjective).[5] Manifested reality is the historical physical universe. It includes all that is or has been accessible to the senses, the present as well as the past, but it excludes everything that we call the future. Manifesting or subjective reality is the future and the mental. It lies in the realm of expectancy and of desire. It is that which is about to be manifest, but which is not fully operating. Space takes both subjective and objective forms. Subjective space belongs to the mental realm: it signifies the heart of things, the "inner" aspect of experience, and it is symbolized by the vertical axis pointing to zenith and the underworld. Objective space radiates from each subjective axis and is essentially the horizontal plane oriented to the four cardinal directions. Cyclical time—the movements of the sun and the pendulumlike swing of the seasons—is located on objective space. To an agricultural people like the Hopi it is important to mark the positions of sunrise and sunset, shifting in the course of the year, on the circumambient horizon.

Distance belongs to the objective realm. The Hopi do not abstract time from distance, and hence the question of simultaneity is to them an unreal problem. They do not ask whether the events in a distant village occur at the same time as those in one's own village. What happens in a distant village can be

Time in Experiential Space

HOPI SPACE AND TIME: SUBJECTIVE AND OBJECTIVE REALMS

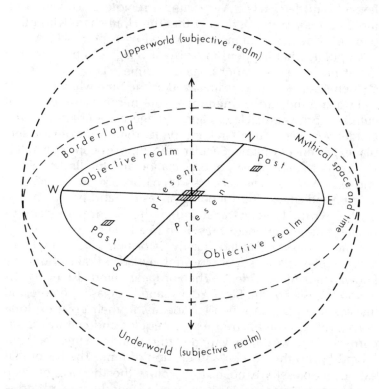

Figure 15. Hopi space and time: subjective and objective realms. The objective realm is the horizontal space within the cardinal grid, but at the distant edges it merges with the subjective realm as represented by the vertical axis.

known here only later. The greater the distance the greater the lapse of time, and the less certain one can be of what has happened out there. Thus distance, although it belongs to the objective realm, does so within limits. As the objective horizontal plane stretches away from the observer to the remote distance, a point is reached at which details cease to be knowable. This is the borderland between the objective and the subjective realms; it is the timeless past, a country told about in myths (Fig. 15).

Time in Experiential Space

"Long ago and far away" are the opening words of many legends and fairy tales. Associating a remote place with a remote past is a way of thinking that the Hopi share with other peoples. The association is supported by experience. As the Hopi put it, what happens in a distant village can be known to me at my place only after a lapse of time. "Long ago" is the Golden Age.[6] Antiquity is idealized as the time when the gods still walked the earth, when men were heroes and bearers of culture, and when sickness and old age were unknown. The Golden Age is shrouded in mystery, far removed from the secular experiences of time. Secular time imposes constraints. It is felt as alternating phases of expectation and fulfillment, effort and ease. Secular time accords with nature's short and observable periodicities. The founding ancestors and heroes of the mythic world largely transcended these normal cycles of human time experience; they lived in a timeless past.

Timelessness is another quality of distant places. In Taoist lore, timeless paradises are located myriads of miles from any known human settlement.[7] The European mind also envisions atemporal Isles of the Blest, Edens, and Utopias in remote and inaccessible places.[8] When Europeans, in their great explorations overseas, discovered exotic peoples and cultures in far corners of the world, they tended to romanticize them and put them beyond the burden and erosion of time. The belief that exotic peoples have no history colors the thinking of even modern ethnographers. The absence of a written record and of ruins signifying the stages of the past may have encouraged this belief. On the other hand ethnographers, like other people, may simply be predisposed to associate the distant with the timeless. The desire common among vacationers to go as far away as possible from their homes is also suggestive. Remote resorts are removed from the burdens of time: in Hopi terms they lie at the borderland between objective and subjective realms. Viewed from the familiar home, they are almost mythic places.

Space is historical if it has direction or a privileged perspective. Maps are ahistorical, landscape paintings are historical.

Time in Experiential Space

The map is God's view of the world since its sightlines are parallel and extend to infinity; orthographic map projection dates back to the ancient Greeks. The landscape picture, with its objects organized around a focal point of converging sightlines, is much closer to the human way of looking at the world; yet it appeared in Europe only in the fifteenth century. Since then landscape pictures that transform "the simultaneity of space into a happening in time—that is, an irreversible sequence of events" have become increasingly popular.[9] Seeing landscape in perspective presupposes a major reordering of time as well as of space. From the Renaissance onward, time in Europe was steadily losing its repetitious and cyclical character and becoming more and more directional. The image of time as swinging pendulum or as circular orbit ceded to the image of time as arrow. Space and time have gained subjectivity by being oriented to man. Of course space and time have always been structured to conform with individual human feelings and needs; but in Europe this fact rose closer to the surface of consciousness at a certain period of its history and found expression in art. Photography in the last hundred years has strengthened and popularized the perspectival vision. Anyone with a simple box camera can now produce an image that melds time into pace.

Under the influence of landscape pictures, painted or captured by the camera, we learn to organize visual elements into a dramatic spatio-temporal structure. When we look at a country scene we almost automatically arrange its components so that they are disposed around the road that disappears into the distant horizon. Again, almost automatically we imagine ourselves traveling down that road; its converging borders are like an arrow pointing to the horizon, which is our destination and future. The horizon is a common image of the future. Statues of statesmen are put on high pedestals, and sculptors show the figures gazing farsightedly at the horizon. Open space itself is an image of hopeful time.[10] Open space is cone-shaped: it opens up from the point where one stands, to the broad horizon that separates earth from sky. Many American homes have large picture windows. A guest, upon entering his host's home, may go straight to the picture window and admire what lies far beyond the house. The host is not displeased. After all, the guest is admiring his prospect, and prospect means both a broad view and future promise. A traditional Oriental home, in contrast, has no picture windows; the rooms look inward to the interior courtyard, and the only expanse of nature visible to the inhabitants is the overarching sky. The vertical axis, rather than the open horizontal space, is the symbol of hope.

The road and its sentinel of trees converge to a vanishing point: it is the path down which we shall travel. Every perspective landscape painting or photograph teaches us to see time "flowing" through space. The distant view need not call forth the idea of future time; the view could be our backward glance and the vanishing road the path we have already trodden. Both the past and the future can be evoked by the distant scene. By the fifteenth century the rules of perspective painting were mastered; landscapes opened up distant vistas. In the course of the eighteenth century background objects in a scene became so important that their treatment tended to prevail even over objects in the foreground. Eighteenth-century aestheticism required that one's eyes be directed outward to the distant scene where one's mind could rest and find personal meaning in the past, future, or eternity.[11] Time, symbolized by far objects in a person's present visual field, could be contemplated. Here is an example from John Dyer's "Grongar Hill" (1726). Look at the prospect, the poet says, and it reveals a future that may deceive.

> How close and small the hedges lie!
> What streaks of meadow cross the eye!
> A step, methinks, may pass the stream,
> So little distant dangers seem;
> So we mistake the future's face.
> Ey'd thro' Hope's deluding glass;
> As you summits soft and fair,
> Clad in colours of the air,
> Which to those who journey near,
> Barren, brown, and rough appear.

Time in Experiential Space

Thomas Gray saw the past in the distant scene. In "Ode on a Distant Prospect of Eton College" (1747), the poet reminisced wistfully over youth.

> Ah, happy hills, ah, pleasing shade,
> Ah, fields belov'd in vain,
> Where once my careless childhood stray'd,
> A stranger yet to pain!

When we stand before a prospect, our mind is free to roam. As we move mentally out to space, we also move either backward or forward in time. Physical movement across space can generate similar temporal illusions. When travel brochures tell us to "step into" the past or future, what they intend is that we should visit a historic or futuristic place—a house or city. We are invited to step into an environment that was built in the past or into one that is made in the style of an imaginary future. Even the layman can roughly tell the age of buildings. He knows the difference between a Victorian mansion and a contemporary ranch house, between an old and a new city. When the tourist steps into an old city he feels he has moved back in time. The natural landscape has a far longer past than anything man-made. Although the layman may be insensitive to the age of the natural environment, explorers and geologists can read time in rock formations. They are also attuned to the time-depth of ruins. For some seventy years after the middle of the nineteenth century, European explorers searched for the source of the Nile in Africa and for signs of ancient civilization in the interior of Asia. Narratives of their journeys give the impression of odysseys into the past rather than into the future. Why? One reason may be the common belief in the antiquity of the African and Asian continents. Popular as well as scientific works characterized these broad landmasses as cradles of mankind and of civilization. Africa was antediluvian, its people "pre-adamite"; [12] Asia was a museum of dead cultures. Exploring these places was like visiting a historic city or museum in which every object reminded the visitor of a remote past.

Geological antiquity and human ruins contributed to the sense of temporal depth, but other psychological dispositions and impulses seemed to be at work. They can, perhaps, be described this way. When we look outward we look at the present or future; when we look inward (that is, introspect) we are likely to reminisce the past. "Inland," the secluded landscape, was for Wordsworth both an image of nature's time rising out of the mists of antiquity and of the remembered past of human time.[13] "Inland," "source," "center," or "core"— these symbols of the exploration mystique—all convey the idea of beginning and of past time. Going up a river to its source is to return, symbolically, to the beginning of one's own life; and in the case of the Nile, to the birthplace of mankind. "Center" means also "origin" and carries a sense of starting point and beginning. Hence, although exploration for the explorer is a thrust into virgin territory, he experiences the move into the heart of a continent as a return to ancient roots, to a country once known but long since forgotten. In addition to Africa and Asia the core of Australia is, in the reflective imagination, veiled in the dust of antiquity.[14] To move into the core of Australia is to go back in time. What of North America? Americans had to concede Europe's older architectural heritage; however, they could and did boast of their land's geological age.[15] For some travelers geological ruins, so prominent in the arid West, assumed the human import and aesthetic values of architectural ruins. Despite these hints of age, emigrants who went west into North America's core surely did not feel they were plunging into the past; on the contrary, they probably felt they were moving onto virgin land and into a spacious future.

Space has temporal meaning in the reflections of a poet, in the mystique of exploration, and in the drama of migration. Space also has temporal meaning at the level of day-to-day personal experiences. Language itself reveals the intimate connectivity among people, space, and time. I am (or we are) here; here is now. You (or they) are there; there is then, and then refers to a time which may be either the past or the future. "What happens then?" The "then" is the future. "It was cheaper then." The "then" here is the past. *Einst*, a German word, means "once," "once upon a time," and "some day (in the future)." Personal pronouns are tied not only to spatial demonstratives (this, that, here, there), but also to the adverbs of time "now" and "then." Here implies there, now implies then. "Implies," however, is a weak verb. Here does not entail there, nor now then. As Thomas Merton put it, life may be so cool that "here" does not even warm itself up with references to "there." The hermit's life is that cool. "It is a life of low definition in which there is little to decide, in which there are few transactions or none, in which there are no packages delivered."[16]

In purposeful activity space and time become oriented with respect to the cogitating and active self. Are not most human activities purposeful? Objectively, yes, for such movements as brushing one's teeth and going to the office can clearly be understood in terms of ends and means. Subjectively even complicated repetitive movements turn into habit; their original intentional structure—envisaging ends and the means to achieve them—is lost. It is only when we reflect on commonplace activities that their original intentional structures reemerge. And of course when we make new plans space and time rise to the surface of consciousness and become aligned with respect to goals.

Consider the routine of going to work in the morning and returning home at night. A man is so habituated to this circuit that he makes it with little deliberation. One does not celebrate a routine. Work at the office promises little excitement; it will be just another day. Yet a ritual surrounds these to-and-fros. Going to the office remains a small adventure. Husbands are "sent off." Each day is a new day. In the morning the office lies ahead, in one's future. Movement to it is forward movement. Office work may be dull most of the time, but novelty is always a possibility if only because one is likely to meet strangers whose behavior cannot be predicted. Uncertainty and the potential for surprise are characteristics of the future and contribute to a sense of the future. At the end of the day the office worker puts on his coat and prepares to return home. Home is now in his future in the sense that it takes time to get there, but he is not likely to feel that the return journey is a forward movement in time. He returns—tracing his steps back in space and going back in time—to the familiar haven of the home. Familiarity is a characteristic of the past. The home provides an image of the past. Moreover in an ideal sense home lies at the center of one's life, and center (we have seen) connotes origin and beginning.

Time and space are directed when one is actively planning. Plans have goals. Goal is a temporal as well as spatial term. European emigrants have a definite goal, which is to settle in the New World. The New World is a place across the Atlantic Ocean. It shares present space with Europe. Yet it is the *New* World, a beckoning future to emigrants. Plans need not be so grand as emigration to a new continent to add a temporal dimension to oriented space. Any effort to envisage a goal—a different resort for the family's next vacation, for example—generates a spatio-temporal structure. Habit, by dulling the sense of purpose and of anxious striving, weakens it. And this sense can be abrogated by leading a cool life, "a life of low definition in which there is little to decide," as Merton said of the hermit.

Music can negate a person's awareness of directional time and space. Rhythmic sound that synchronizes with body movement cancels one's sense of purposeful action, of moving through historical space and time toward a goal. Walking purposefully from A to B is felt as leaving so many steps behind and as having so much more ground ahead to cover. Change the environment by introducing band music and, objectively, one still marches from A to B with seeming deliberation. Subjectively, however, space and time have lost their directional thrust under the influence of rhythmic sound. Each step is no longer just another move along the narrow path to a destination; rather it is striding into open and undifferentiated space. The idea of a precisely located goal loses relevance.

Normally a person feels comfortable and natural only when he steps forward. Stepping back feels awkward, and one remains apprehensive even when assured that nothing lies behind to cause a stumble. Dancing, which is always accompanied by music or a beat of some kind, dramatically abrogates historical time and oriented space. When people dance they

Time in Experiential Space

move forward, sideways, and even backward with ease. Music and dance free people from the demands of purposeful goal-directed life, allowing them to live briefly in what Erwin Straus calls "presentic" unoriented space.[17] Soldiers who march to military music tend to forget not only their weariness but also their goal—the battlefield, with its promise of death. In modern society Musak in offices and shopping centers and pulsating sound from transistor radios for teenagers suggest that people want to forget a space-time frame tied to goals, many of which are perceived to be unattractive or meaningless.

Historical time and oriented space are aspects of a single experience. Intention creates a spatio-temporal structure of "here is now," "there is then." I can reflect on this structure and say, here is point A and there is point B: what is the distance between them? Having recognized that village B is my goal—a point out there in space and my future—the pragmatic question arises, how far is it from me? The answer is frequently given in units of time: village B is two "sleeps," or two days, away; it is a half-hour's drive. Here then is another relation between distance and time—time as a measure of distance. For purposes of measurement, time is not envisioned as an arrow pointing to the future; rather time is perceived to be repetitious, like the swing of the pendulum, and it is calibrated to internal biological rhythms as well as to the observable periodicities of nature.

An explanation for the wide use of time to measure distance is the fact that units of time convey a clear sense of effort. The useful answer to questions of distance tells us how much effort is needed—what resources of energy are required—to achieve a goal. The next goal is a spear throw or an arrow shot away; it is one hundred paces away. A reply of this kind appeals directly to experience. Not only are we able to envisage the distance of a spear throw, we can also feel it in the effort of throwing a spear. A pace is not only something we can see—the span between one foot and another—but it is also felt in the muscles. How is the pace (step) or spear throw related to time? A pace is a unit of time because it is felt as a biological arc of effort and ease, strain and relaxation. One hundred paces means one hundred units of a biological rhythm that we know intimately. Another biological rhythm that we know intimately is the cycle of wakefulness and sleep. Distances may be given in "sleeps" or days. The pace, we have noted, is a felt effort as well as a measure that can be seen. Similarly the cycle of wakefulness and sleep is not only the strain of activity followed by rest, but it is visible in external nature as light and darkness and as the trajectory of the sun. Giving distance in days relates to effort in another, more calculating, sense. When we are told that it is three days to another village, we know roughly how much food and water to bring; we can calculate the amount of energy required to take us to our destination. How far is it from Minneapolis to Los Angeles? An answer in miles or kilometers is not very helpful unless these units of distance can be quickly translated into time, effort, and needed resources. In contrast, the answer, "it is a three-day drive," tells us more directly how much money to bring for motels, gas, and food—the money required to purchase energy.

The intention to go to a place creates historical time: the place is a goal in the future. The future cannot, however, be left open and undefined. Emigrants who propose to settle in the interior of the United States must plan to reach their destination at a propitious time, say, spring. Just as the goal is not the United States but a more or less specific area within the United States, so the future is not open-ended time but a particular year and a particular season within the year. This constraint on the future, on historical time, is itself strong reason for estimating distance in time units. The need to be at a certain place nearly always means being there at a certain time. The herdsman has to drive his flock to a particular pasture by a certain date, and the businessman must attend a sales convention in another part of the city by a certain hour. Time everywhere regulates human lives and livelihood. The essential difference between technological and nontechnological societies is that in the former, time is calibrated to the precision of the hour and the minute.[18]

We have seen how space and time coexist, intermesh, and define each other in personal experience. Every activity generates a particular spatio-temporal structure, but this structure seldom thrusts to the front of awareness. Events such as going to the office, planning a visit, looking at the scenery, hearing news of friends in another town, are too much the accepted pattern of daily living to warrant reflective thought. What compels us to reflect on experience? Untoward events. In non-technological societies the forces of nature often seem unpredictable: they are the untoward events that intrude on human lives and command attention. They can be "tamed" by being made a part of a cosmology or world view. Not all cultures have an articulated world view. Where one does exist, relatively few people are capable of conceptualizing it in detail and in a systematic way. World view is at a distance from particular experiences and needs; it is an intellectual construct. In such a construct, we may now ask, how are space and time represented? Does one type of mythic time entail one type of mythic space, and vice versa?

Mythic space is commonly arranged around a coordinate system of cardinal points and a central vertical axis. This construct may be called cosmic, for its frame is defined by events in the cosmos. Mythic time is of three principal kinds: cosmogonic, astronomic, and human. Cosmogonic time is the story of origins, including the creation of the universe. Human time is the course of human life. Both are linear and one-directional. Astronomic time is experienced as the sun's daily round and the parade of seasons; its nature is repetition. Wherever cosmic space is prominently articulated, cosmogonic time tends to be either ignored or weakly symbolized. In North America a common cosmogonic motif among the Indians is that of the earth-diver, who brings up earth from the ocean, creating an island that grows steadily in size. This creation story, unlike cyclical astronomic time, finds no representation in cosmic space. Another type of origin myth is concerned with the birth and achievement of ancestors and culture heroes. The Pueblo Indians of the American Southwest believe that their ancestors emerged from the ground at Shipap in the north and drifted south looking for the "middle place." Elements of this story have left their mark on the Pueblo Indian's cosmic space.

Shipap is located in the north; to the south, midway from the cosmic center, is the White House where the ancestral people paused to acquire cultural skills.[19] The symmetry of Puebloan cosmic space, stretched over a grid of cardinal points, is thus distorted by time's arrow embodying the myth of migration.

The mythic space of Australian natives is not geometrical. Cyclical astronomic time is alien to aboriginal thought. Cosmogonic time, however, is recognized. It leaves its mark on space, thereby sanctioning it. Cosmogonic time in aboriginal thought is not concerned with the creation of land, sky, or sea; the fundamental setting is assumed. Origin myth relates the way the founding ancestors prepared the earth for human habitation, how they supplied the natural resources and changed the landscape by their actions. To Australian natives topographical features are a record of "who were here, and did what." They are also a record of "who are here now."[20] The landscape—though unmodified to Western eyes—documents the achievements of a people. In central Australia and Arnhem Land myths tell how founding ancestors wandered toward distant goals. Sometimes they strayed, and fate would overtake them. Where they died they left their spirits. The landscape is littered with memorials to mythic heroes who were stopped short in their tracks.[21] By conceiving of the heroes as wanderers and by periodically reenacting the mythic journeys, Australian aborigines impart a directional time sense to their land (Fig. 16).

In comparison with cosmogonic time, astronomic time is easily mapped onto a spatial frame. Astronomic time, being cyclical and repetitious, is best represented by symmetrical space. Symmetrical space is a cosmic clock registering the path of the sun. We have noted that cosmogonic time mars this spatial symmetry. Human time is also directional. A human life begins at birth and ends in death: it is a one-way journey. Human time is biased in favor of the future. Life is lived in the future, which may be as close as the next meal or as distant as the next promotion up the ladder of success. Human time, like the human body, is asymmetrical: one's back is to the past, one's face to the future. Living is a perpetual stepping forward

MYTHOLOGICAL TRACKS OF THE WALBIRI

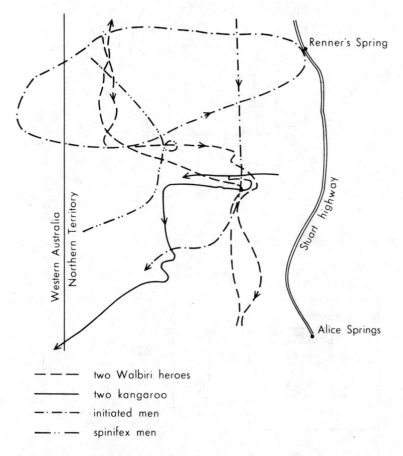

— — — — two Walbiri heroes

———— two kangaroo

—·—··— initiated men

—··— spinifex men

Figure 16. Cosmogonic myth and directed space: mythological tracks of the ancestral heroes of the Walbiri in central Australia. Adapted from Amos Rapoport, "Australian aborigines and the definition of place," in W. J. Mitchell, ed., *Environmental Design and Research Association*, Proceedings of the Third Conference at Los Angeles, 1972, page 3-3-9, figure 4.

A. Symmetrical sacred space

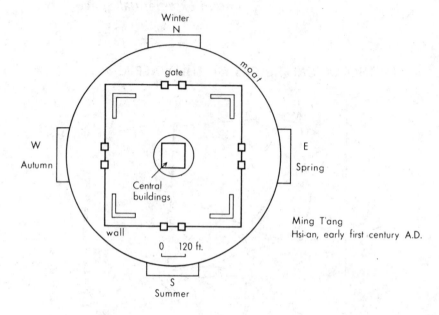

Ming T'ang
Hsi-an, early first century A.D.

B. City of Man: biased space

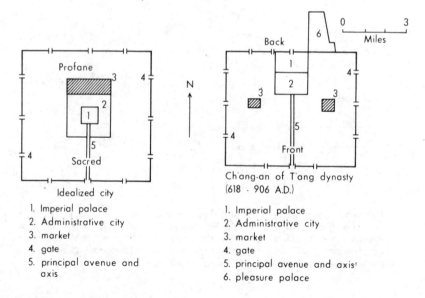

Idealized city

1. Imperial palace
2. Administrative city
3. market
4. gate
5. principal avenue and axis

Ch'ang-an of T'ang dynasty
(618 - 906 A.D.)

1. Imperial palace
2. Administrative city
3. market
4. gate
5. principal avenue and axis
6. pleasure palace

Figure 17. Symmetrical sacred space (Ming T'ang) and asymmetrical or biased space of the City of Man both in its idealized form and in actuality (Ch'ang-an).

into light and abandoning what is behind one's back, cannot be seen, is dark and one's past. Among Pueblo Indians the dead are believed to return to Shipap, from which the ancestors had emerged. More often, however, death is a continuing journey from the center of the cosmic space, either along the vertical axis or to one of the cardinal points. In the world of the living the future and the front are favored; the symmetry of cosmic space is distorted by having a privileged axis and direction. Consider the traditional Chinese city. Its cosmic plan registers the movement of the stars and the march of the seasons. Ideally it should be symmetrical, but it is not; it has a privileged axis, which is the central avenue aligned from the palace southward to the meridian gate. In the capital city the ruler in his palace has his back to the north, to dark and profane space, and he faces south to the world of light and of man. The traditional city, though strongly influenced by cosmic ideas in basic design, is nonetheless (as Nelson Wu puts it) the City of Man.[22] It mirrors the asymmetry of human time and life (Fig. 17).

10

Intimate Experiences
of Place

t is impossible to discuss experiential space without introducing the objects and places that define space. An infant's
space expands and becomes better articulated as he recognizes and reaches out to more permanent objects and places.
Space is transformed into place as it acquires definition and
meaning. We have noted how strange space turns into
neighborhood, and how the attempt to impose a spatial order
by means of a grid of cardinal directions results in the establishment of a pattern of significant places, including the cardinal points and center. Distance is a meaningless spatial concept apart from the idea of goal or place. It is possible, however, to describe place without introducing explicitly spatial
concepts. "Here" does not necessarily entail "there." We can
focus on the experiencing of the "here," and we shall do so in
this and the next two chapters. We move from direct and intimate experiences to those that involve more and more symbolic and conceptual apprehension.

Intimate experiences lie buried in our innermost being so
that not only do we lack the words to give them form but often
we are not even aware of them. When, for some reason, they
flash to the surface of our consciousness they evince a poignancy that the more deliberative acts—the actively sought

experiences—cannot match. Intimate experiences are hard to express. A mere smile or touch may signal our consciousness of an important occasion. Insofar as these gestures can be observed they are public. They are also fleeting, however, and their meaning so eludes confident interpretation that they cannot provide the basis for group planning and action. They lack the firmness and objectivity of words and pictures.

Intimate occasions are often those on which we become passive and allow ourselves to be vulnerable, exposed to the caress and sting of new experience. Children relate to people and objects with a directness and intimacy that are the envy of adults bruised by life. Children know they are frail; they seek security and yet remain open to the world. In sickness adults also know frailty and dependency. A sick person, secure in the familiarity of his home and comforted by the presence of those he loves, appreciates the full meaning of nurture. Intimate places are places of nurture where our fundamental needs are heeded and cared for without fuss. Even the vigorous adult has fleeting moments of longing for the kind of coziness he knew in childhood. What sensual ease compares with that of a child as he rests in the parent's arm and is read to sleep? In the curve of the human arm is comfort and security absolute, made all the more delectable by the threatening wolf in the storybook. As adults, after a day of strenuous assertion, we sink gratefully into an armchair and relax in its receptive hollow while we watch televised news of mayhem. The home itself feels more intimate in winter than in summer. Winter reminds us of our vulnerability and defines the home as shelter.[1] Summer, in contrast, turns the whole world into Eden, so that no corner is more protective than another.

Unique to human beings among primates is the sense of the home as a place where the sick and the injured can recover under solicitous care. Washburn and De Vore, in their account of the society of early man, note that all human societies have bases where the weak may stay and from which the fit may move out to gather, hunt, or fight. In the home base are tools, food, and normally some sort of shelter. "No such 'base' exists among the baboons, other monkeys, or apes. When the troop

moves out on the daily round, *all* members must move with it
or be deserted. . . . [T]he only protection for a baboon is to
stay with the troop, no matter how injured or sick he may
be. . . . For a wild primate a fatal sickness is one that separates it from the troop, but for man it is one from which he
cannot recover even while protected and fed at the home
base."[2]

Several conditions necessary for an elemental sense of place
are encapsulated in this brief account. Place is a pause in
movement. Animals, including human beings, pause at a locality because it satisfies certain biological needs. The pause
makes it possible for a locality to become a center of felt value.
Baboons and apes apparently do not pause in order to take
care of an injured or sick member. Humans do, and this fact
contributes to the depth of their sentiment for place. A person
recovering from sickness is aware of his dependence on
others. He is aware that he is cared for and made well at a
specific locale, which may be the shade of a tree, a lean-to
shelter, or a fourposter bed. At one place the patient is cradled
back to health. Before full recovery he remains for a time weak
and passive like a child; he is able to respond to the immediacy
of the world and see it with the fresh intensity of childlike eyes.
The lasting affection for home is at least partly a result of such
intimate and nurturing experiences.

To the young child the parent is his primary "place." The
caring adult is for him a source of nurture and a haven of
stability. The adult is also the guarantor of meaning to the
child, for whom the world can often seem baffling. A mature
person depends less on other people. He can find security and
nourishment in objects, localities, and even in the pursuit of
ideas. Home, for the conductor Bruno Walter, was the world of
classical music. He did not feel estranged when he had to
abandon his native Austria for the United States. Exceptionally
talented people can live for art or science and go wherever
they thrive. There are also recluses and misanthropes who
shun men in favor of the consolation that nature or material
possessions can provide.[3] For most people possessions and
ideas are important, but other human beings remain the focus

of value and the source of meaning. We say of young lovers that they dwell in each other's gaze. They are free of attachment to things and to locality; they will abandon their homes and elope if they have to. Old couples are attached to place but they are even more attached to people, services, and each other. Old people may not wish to survive their partners' deaths for long, even when the material conditions that maintain their lives are kept up. For such reasons we speak of *resting* in another's strength and *dwelling* in another's love. Even so, the idea of a human person as "place" or "home" is not immediately acceptable.

Tennessee Williams, in a play, suggests how home may well be another person, that is to say, how one human being can "nest" in another. Hannah Jelkes, a middle-aged spinster, and her ancient grandfather are rootless people. They roam the country and try to make a living by selling their fragile skills, she as a quick sketch artist and he as "the world's oldest working poet." The following dialogue is between Hannah and a cynical dissipated man called Shannon. They are on the porch of a run-down hotel in Mexico.

Hannah: We make a home for each other, my grandfather and I. Do you know what I mean by a home? I don't mean a regular home. I mean I don't mean what other people mean when they speak of a home, because I don't regard a home as a . . . well, as a place, a building . . . a house . . . of wood, bricks, stone. I think of home as being a thing that two people have between them in which each can . . . well, nest—rest—live in, emotionally speaking. Does that make any sense to you, Mr. Shannon?

Shannon: Yeah, complete. But . . . when a bird builds a nest to rest in and live in, it doesn't build it in a . . . falling-down tree.

Hannah: I am not a bird, Mr. Shannon.

Shannon: I was making an analogy, Miss Jelkes.

Hannah: I thought you were making yourself another rum-coco, Mr. Shannon.

Shannon: Both. When a bird builds a nest, it builds it with an eye for . . . the relative permanence of location, and also for the purpose of mating and propagating its species.

Hannah: I still say that I'm not a bird, Mr. Shannon. I am a human being and when a member of that fantastic species builds a nest in the heart of another, the question of permanence isn't the first or even the last thing that's considered . . . necessarily? . . . always?[4]

The dialogue ends on a note of doubt. Permanence is an important element in the idea of place. Things and objects endure and are dependable in ways that human beings, with their biological weaknesses and shifting moods, do not endure and are not dependable. Yet Hannah makes a point. In the absence of the right people, things and places are quickly drained of meaning so that their lastingness is an irritation rather than a comfort. Saint Augustine's native city, Thagaste, was transformed for him with the death of his childhood friend. The great theologian wrote: "My heart was now darkened by grief, and everywhere I looked I saw death. My native haunts became a scene of torture to me, and my own home a misery. Without him everything we had done together turned into excruciating ordeal. My eyes kept looking for him without finding him. I hated all the places where we used to meet, because they could no longer say to me, 'Look, here he comes,' as they once did."[5]

For Augustine the value of place was borrowed from the intimacy of a particular human relationship; place itself offered little outside the human bond. Experiences like that of Augustine are not uncommon. Here is an example from current sociological research. Neilson is a widower. His wife died in childbirth—the birth of her sixth child. Neilson worked as maintenance man for a large firm. By working on the second shift he could see his children off to school and be at home when they returned in the early afternoon. Neilson's younger unmarried sister had moved in with him after his wife died. She would come home about five, cook dinner, put the children to bed, and then retire herself. She was asleep when Neilson returned from work. Neilson returned to a full house but he felt its emptiness. "When I come home from work in the nights," he said, "I feel empty. I feel, coming home, I feel kind of funny, a funny feeling that I'm going into an empty house. Even though the house is still full with the kids, it's just not the same."[6]

Intimate Experiences of Place

Intimacy between persons does not require knowing the details of each other's life; it glows in moments of true awareness
and exchange. Each intimate exchange has a locale which par-
takes in the quality of the human encounter. There are as many
intimate places as there are occasions when human beings
truly connect. What are such places like? They are elusive and
personal. They may be etched in the deep recesses of memory
and yield intense satisfaction with each recall, but they are not
recorded like snapshots in the family album, nor perceived as
general symbols like fireplace, chair, bed, and living room that
invite intricate explication. One can no more deliberately design such places than one can plan, with any guarantee of
success, the occasions of genuine human exchange. Consider
the following description of a brief encounter and its setting;
neither is so unusual that it calls for special notice, yet they are
the stuff that enrich people's lives. In a novel by Christopher
Isherwood, George is a lecturer at a California state college. As
George steps out of the classroom building, the first people he
recognizes are two of his favorite students, Kenny Potter and
Lois Yamaguchi.

They are sitting on the grass under one of the newly planted trees.
Their tree is even smaller than the others. It has barely a dozen leaves
on it. To sit under it at all seems ridiculous; perhaps this is just why
Kenny chose it. He and Lois look as though they were children playing
at being stranded on a South Pacific atoll. Thinking this, George smiles
at them. They smile back. . . . George passes quite close by their
atoll as a steamship might, without stopping. Lois seems to know what
he is, for she waves gaily to him exactly as one waves to a steamship,
with an enchantingly delicate gesture of her tiny wrist and hand.
Kenny waves also, but it is doubtful if *he* knows; he is only following
Lois's example. Anyhow, their waving charms George's heart. He
waves back to them. The old steamship and the young castaways have
exchanged signals—but not signals for help. . . . Again, as by the
tennis players, George feels that his day has been brightened.[7]

Trees are planted on campus to give it more shade and to
make it look greener, more pleasant. They are part of a deliberate design to create place. Having only a few leaves, the trees
do not yet make much of an aesthetic impact. Already, however, they can provide a stage for warm human encounters;

each sapling is a potential place for intimacy, but its use cannot be predicted since this depends on chance and on the play of imagination.

What things move us? What is the most beautiful thing in Belvedere? Belvedere, in Paul Horgan's novel, is the name of a small town on the west central plains of Texas. A teenager in the novel poses the question and answers:

It's not what they brag about, the lilacs, and the green tile dome on the city hall, and the Greek pillars on the bank. No, it is what happens after the sun goes down, and the vapor lights on the tall aluminum poles over the highway start to come on! Do you think I am raving? . . . You know: the sky is still brilliant, but evening is coming, and for the first five minutes or so, the vapor lamps have a color . . . and the thing is so magic when it happens it is enough to make you dizzy. Everything on the earth is sort of gray by then, yes, lilac gray, and there are shadows down the streets, but there, while the sky is changing, those lights are the most beautiful things in the United States! And you know? It's all an accident! They don't *know* how beautiful the light is.[8]

Different things move us. In a short story John Updike makes his hero, David Kern, say:

I, David Kern, am always affected—reassured, nostalgically pleased, even, as a member of my animal species, made proud—by the sight of bare earth that has been smoothed and packed firm by the passage of human feet. Such spots abound in small towns: the furtive break in the playground fence dignified into a thoroughfare, the trough of dust underneath each swing . . . the blurred path worn across a wedge of grass, the anonymous little mound or embankment polished by play and strewn with pebbles like the confetti aftermath of a wedding. Such unconsciously humanized intervals of day, too humble and common even to have a name, remind me of my childhood, when one communes with dirt down among the legs, as it were, of presiding fatherly presences. The earth is our playmate then, and the call to supper has a piercingly sweet eschatological ring.[9]

The modest work of human erosion, Updike continued, "seemed precious because it had been achieved accidentally, and had about it that repose of grace that is beyond willing." Accident and happy chance, these are key ideas in the three examples taken from the works of Isherwood, Horgan, and Updike. Trees are planted for aesthetic effect, deliberately, but their real value may lie as stations for poignant, unplanned human encounters. Highway lamps are functional, yet at sundown their vapor lights can produce colors of dizzying beauty, "the most beautiful things in the United States." The trough of dust under the swing and the bare earth packed firm by human feet are not planned, but they can be touching. Intimate experiences, not being dressed up, easily escape our attention. At the time we do not say "this is it," as we do when we admire objects of conspicuous or certified beauty. It is only in reflection that we recognize their worth. At the time we are not aware of any drama; we do not know that the seeds of lasting sentiment are being planted.

Humble events can in time build up a strong sentiment for place. What are these events like and how do they depend on the feel of things? On a warm May day in one of the hollows of Appalachia a child had just been breast-fed. Robert Coles, observing life in the hollows, noted how suddenly the mother put the child down on the ground, and gently fondled him and moved him with her bare feet. She spoke gravely to her child: "This is your land, and it's about time you started getting to know it."[10] Another mother said to Coles: "When one of my kids starts getting all teary, and there's something bothering him, you know—then is the time for me to help as best I can; and there's nothing that'll work better than getting a child to see if the chickens have laid any new eggs, or to count how many tomatoes there are hanging on the plants, ready for us to pick."[11]

Chickens, eggs, and tomatoes are commonplace objects on the farm. They are there to be eaten or marketed; they are not aesthetic objects. Yet they seem to have at times the essence of wholesome beauty, and they can console. The contemplation and handling of a jug or a warm but firm tomato can somehow reassure us, in depressed moods, of the ultimate sanity of life. In Doris Lessing's novel *The Golden Notebook*, Anna felt that an unpleasantly grinning man was following her. She wanted to run. Panic threatened to engulf her although she knew that her fear was largely irrational.

She thought: if I could see something or touch something that wasn't ugly. . . . There was a fruit barrow just ahead, offering tidy coloured loads of plums, peaches, apricots. Anna bought fruit: smelling at the tart clean smell, touching the smooth or faintly hairy skins. She was better. The panic had gone. The man who had been following her stood near, waiting and grinning; but now she was immune from him. She walked passed him immune.[12]

The home place is full of ordinary objects. We know them through use; we do not attend to them as we do to works of art. They are almost a part of ourselves, too close to be seen. Contemplate them and what happens? Nausea, for the lacerated sensibility of Sartrean man. For Wright Morris the word "holiness" comes to mind. He asked: "Was there, then, something holy about these things? If not, why had I used that word? For holy things, they were ugly enough." Morris looked at the odds and ends on the bureau, the pin-cushion lid on the cigar box, the faded Legion poppies, assorted pills, patent medicines, and concluded that "there was not a thing of beauty, a man-made loneliness anywhere." Yet he was feeling, at that moment, what he expected a thing of beauty could make him feel—an independent presence. People dare not feel for long, Morris asserted. To keep feeling at bay we call on embarrassment. Embarrassment "snaps it off, like an antisepsis, or we rely on our wives, or one of our friends, to take the pressure out of the room with a crack of some kind."[13]

Home is an intimate place. We *think* of the house as home and place, but enchanted images of the past are evoked not so much by the entire building, which can only be seen, as by its components and furnishings, which can be touched and smelled as well: the attic and the cellar, the fireplace and the bay window, the hidden corners, a stool, a gilded mirror, a chipped shell. "In smaller, more familiar things," says Freya Stark, "memory weaves her strongest enchantments, holding us at her mercy with some trifle, some echo, a tone of voice, a scent of tar and seaweed on the quay. . . . This surely is the meaning of home—a place where every day is multiplied by all the days before it."[14]

Hometown is an intimate place. It may be plain, lacking in

architectural distinction and historical glamor, yet we resent an outsider's criticism of it. Its ugliness does not matter; it did not matter when we were children, climbed its trees, paddled our bikes on its cracked pavements, and swam in its pond. How *did* we experience such a small, familiar world, a world inexhaustibly rich in the complication of ordinary life but devoid of features of high imageability? To prompt our memory Helen Santmyer wrote:

You passed the doctor's office, and were at the corner of your own street, where you turned west, and saw the trees arched against a glowing sky. Perhaps you went toward them thinking of nothing much, comfortably aware that you were nearly home. Perhaps, if the skies were gray, if it were winter and the pavement were streaked with soot, and lumps of black snow filled the gutter, you were even remarking how ugly the town was, and how drab and dull. If the skies were clear, you almost certainly paused at the gate, with a hand on the latch, to search for the first star in the west, to wish for escape and a brilliant future far, far away—and yet at the same instant you were aware of the iron of the gate beneath your hand, and were storing away the memory of how it felt.

And so the touch and heart make up their magpie hoard, heedless of the calculating eye and intelligence. "Valentines in a drugstore window, the smell of roasting coffee, sawdust on the butcher's floor—there comes a time in middle age when even the critical mind is almost ready to admit that these are as good to have known and remembered . . . as fair streets and singing towns and classic arcades."[15]

Home place and quotidian life feel real. An Illinois farm girl went with her husband to California for their honeymoon. She said:

We didn't stay as long as we planned; we came right back here. We do that all the time when we take trips; we can't wait to come back. It's so unreal to be gone. That's the unreal world. We know where life begins and ends here. Life goes on here. It's nice to think about going away and doing, getting away from it, but it's always nice to get back to life that really is. When I think about it, it was like a waste of time. Our real life was back here. We wanted to get back and start living.[16]

What does the Illinois farm girl mean by "real"? It is hard to say. The real, we feel, is important, but paradoxically it also goes unnoticed. Life is lived, not a pageant from which we stand aside and observe. The real is the familiar daily round, unobtrusive like breathing. The real involves our whole being, all our senses. On vacation, although problems have been left behind, an important part of ourselves has also been left behind; we become specialized and unanchored beings, sightseers who sample life effortlessly.

Seeing has the effect of putting a distance between self and object. What we see is always "out there." Things too close to us can be handled, smelled, and tasted, but they cannot be seen—at least not clearly. In intimate moments people shade their eyes. Thinking creates distance. Natives are at home, steeped in their place's ambience, but the instant they think about the place it turns into an object of thought "out there." Tourists seek out new places. In a new setting they are forced to see and think without the support of a whole world of known sights, sounds, and smells—largely unacknowledged— that give weight to being: vacation areas, however delightful, seem unreal after a time.

In Santmyer's recollection of her hometown, she contrasts vision with touch. Seeing, like thought, is evaluative, judgmental, and conducive to fantasy. If the sky were gray, she said, you would remark on "how ugly the town was, and how drab and dull." And if the sky were clear, you would pause at the gate to wish for escape and a bright future far away. Images and ideas discharged by the mind are seldom original. Evaluations and judgments tend to be clichés. The fleeting intimacies of direct experience and the true quality of a place often escape notice because the head is packed with shopworn ideas. The data of the senses are pushed under in favor of what one is taught to see and admire. Personal experience yields to socially approved views, which are normally the most obvious and public aspects of an environment. To illustrate, here is Robert Pirsig's account of how tourists see Crater Lake in Oregon:

At the lake we stop and mingle affably with the small crowd of tourists holding cameras and children yelling, "Don't go too close!" and see cars and campers with all different license plates, and see the Crater Lake with a feeling of "Well, there it is," just as the pictures show. I watch the other tourists, all of whom seem to have out-of-place looks too. I have no resentment at all this, just a feeling that it's all unreal and that the quality of the lake is smothered by the fact that it's so pointed to. You point to something as having Quality and the Quality tends to go away. Quality is what you see out of the corner of your eye, and so I look at the lake below but feel the peculiar quality from the chill, almost frigid sunlight behind me, and the almost motionless wind.[17]

Intimate experiences, whether of people or of things, are difficult to make public. Apt words are elusive; pictures and diagrams seldom seem adequate. Music can evoke certain feelings, but it lacks denotative precision. Facts and events are readily told: we have no problem saying that we went to Crater Lake on a Sunday, with the children and two dogs, in a station wagon, and that it was a cold day. We know what to admire: the lake. We can point to it and take a picture so that it stays with us as a permanent and public record of what has happened. But the quality of the place and of our particular encounter are not thus captured: *that* must include what we see out of the corner of our eye and the sensation of the almost frigid sunlight behind us.

Intimate experiences are difficult but not impossible to express. They may be personal and deeply felt, but they are not necessarily solipsistic or eccentric. Hearth, shelter, home or home base are intimate places to human beings everywhere. Their poignancy and significance are the themes of poetry and of much expository prose. Each culture has its own symbols of intimacy, widely recognized by its people. Americans, for example, respond to such emblems of good life as the New England church, the Middle Western town square, the corner drugstore, Main Street, and the village pond.[18] An armchair or a park bench can be an intensely personal place, yet neither is a private symbol with meanings wholly opaque to others.[19] Within a human group experiences have sufficient overlap so that an individual's attachments do not seem egregious and incomprehensible to his peers. Even an experience that appears to be the product of unique circumstances can be shared. The scene drawn by Isherwood, in which a teacher makes brief contact with two students sitting under a newly planted tree on a California campus, is highly specific. Its meaning, however, is not impenetrably private: all who read the passage and nod in recognition, whether or not they have taught in an American college or lived in California, share it to some degree.

There is far more to experience than those elements we choose to attend to. In large measure, culture dictates the focus and range of our awareness. Languages differ in their capacity to articulate areas of experience. Pictorial art and rituals supplement language by depicting areas of experience that words fail to frame; their use and effectiveness again vary from people to people. Art makes images of feeling so that feeling is accessible to contemplation and thought. Social chatter and formulaic communication, in contrast, numb sensitivity. Even intimate feelings are more capable of being represented than most people realize. The images of place, here sampled, are evoked by the imagination of perceptive writers. By the light of their art we are privileged to savor experiences that would otherwise have faded beyond recall. Here is a seeming paradox: thought creates distance and destroys the immediacy of direct experience, yet it is by thoughtful reflection that the elusive moments of the past draw near to us in present reality and gain a measure of permanence.

11

Attachment to Homeland

Place exists at different scales. At one extreme a favorite armchair is a place, at the other extreme the whole earth. Homeland is an important type of place at the medium scale. It is a region (city or countryside) large enough to support a people's livelihood. Attachment to the homeland can be intense. What is the character of this sentiment? What experiences and conditions promote it?

Human groups nearly everywhere tend to regard their own homeland as the center of the world. A people who believe they are at the center claim, implicitly, the ineluctable worth of their location. In diverse parts of the world this sense of centrality is made explicit by a geometrical conception of space oriented to the cardinal points. Home is at the center of an astronomically determined spatial system. A vertical axis, linking heaven to the underworld, passes through it. The stars are perceived to move around one's abode; home is the focal point of a cosmic structure. Such a conception of place ought to give it supreme value; to abandon it would be hard to imagine. Should destruction occur we may reasonably conclude that the people would be thoroughly demoralized, since the ruin of their settlement implies the ruin of their cosmos. Yet this does not necessarily happen. Human beings have strong recuperative powers. Cosmic views can be adjusted to suit new circumstances. With the destruction of one "center of the world," another can be built next to it, or in another location altogether, and it in turn becomes the "center of the world." "Center" is not a particular point on the earth's surface; it is a concept in mythic thought rather than a deeply felt value bound to unique events and locality. In mythic thought several world centers may coexist in the same general area without contradiction. It is possible to believe that the axis of the world passes through the settlement as a whole as well as through the separate dwellings within it. Space that is stretched over a grid of cardinal points makes the idea of place vivid, but it does not make any particular geographical locality *the* place. A spatial frame determined by the stars is anthropocentric rather than place-centric, and it can be moved as human beings themselves move.

If a cosmic world view does not guarantee uniqueness to locality, what beliefs do? Evidence from different cultures suggests that place is specific—tied to a particular cluster of buildings at one location—wherever the people believe it to be not only their home but also the home of their guarding spirits and gods. Ancient cities in the Near East and in the Mediterranean Basin enjoyed this kind of particularity. The original inspiration for building a city was to consort with the gods. Early Mesopotamian towns were essentially temple communities. Ritual centers and the more important settlements in the Nile Valley also had religious foundations, since they were thought to occupy sites on which primordial creation had taken place. It is difficult for the modern mind to appreciate the extent to which religion intermeshed with human activities and values in ancient times. When life seemed uncertain and nature hostile, the divinities not only promoted life and protected it, they were also guarantors of order in nature and in society. The legitimacy of laws and institutions depended on them. The withdrawal of the presiding presences meant chaos and death. Conquerors did not raze a city to the ground simply out of wanton fury; in such destruction they appropriated a people's gods by rendering them homeless, and in appropriating the gods the conquerors acquired a civilization. This belief throws light on the paradox that, although the city is the embodiment of civilization, the Sumerians listed "the destruction of cities" as one of the divine institutions upon which civilization is founded.[1]

In the Mycenaean period Greek cities owed their sacred status to their divine residents. Athena and Helen were Mycenaean goddesses who presided over Athens and Sparta respectively. In these prehistoric times of kingly rule, shrines had an importance they would later lose during the republican period. A Helladic city, however straitened by its enemies, remained viable so long as the shrines housing the divine images were intact. This belief, says John Dunne, "is reflected to some extent in the tradition of the Trojan War according to which it was necessary to steal the Palladium, the image of the city-goddess, from Troy before the city could be taken."[2] Removal of the image, or destruction of the shrine that housed it, would have deprived a city of its legitimacy since the rules, rites, and institutions under which a people lived all required divine sanction. We cannot know prehistoric sentiments: they are at best matters for conjecture. From the historic period of the ancient Mediterranean world we can find many expressions of love for place. One of the most eloquent was attributed to a citizen of Carthage. When the Romans were about to destroy Carthage at the end of the third Punic War, a citizen pleaded with them thus:

We beseech you, in behalf of our ancient city founded by the command of the gods, in behalf of a glory that has become great and a name that has pervaded the whole world, in behalf of the many temples it contains and of its gods who have done you no wrong. Do not deprive them of their nightly festivals, their processions and their solemnities. Deprive not the tombs of the dead, who harm you no more, of their offerings. If you have pity for us . . . spare the city's hearth, spare our forum, spare the goddess who presides over our council, and all else that is dear and precious to the living. . . . We propose an alternative more desirable for us and more glorious for you. Spare the city which has done you no harm, but, if you please, kill us, whom you have ordered to move away. In this way you will seem to vent your wrath upon men, not upon temples, gods, tombs, and an innocent city.[3]

It is true that this plea was written in the second century A.D. several hundred years after the event. How the besieged Carthaginians really felt we have no way of knowing. But the plea at least made good sense to Roman readers, for whom it was written, whereas to us it verges on the incredible. Suppose that Martians have invaded America and are at the gates of Minneapolis. It is hard to believe that our city councilors will plead with the Martians to kill us but save Nicollet Mall, which has done them no harm.

Religion could either bind a people to place or free them from it. The worship of local gods binds a people to place whereas universal religions give freedom. In a universal religion, since all is created by and all is known to an omnipotent and omniscient god, no locality is necessarily more sacred than another. Historically, earthbound deities reigned prior to the appearance of universal sky gods. Perhaps people everywhere have entertained the idea of a universal divinity, but his presence was shadowy and remote in comparison with the local spirits that constantly intruded on human affairs. In China the idea of *t'ien* (heaven) evolved and rose to the fore of consciousness in the course of the Chou dynasty (ca. 1027–256 B.C.). *Ti* (earth) was its counterpart, though of somewhat inferior status. *T'u* or soil gods occupied still lower ranks, but they were primordial. T'ien and ti were sophisticated concepts; in comparison soil gods and the many spirits of nature had much greater reality for the people. In the Mediterranean world the sky gods of Olympus were firmly installed by Homeric times. Men, however, did not at first conceive of these divinities as watching over the whole human race; rather they thought of each as belonging to a particular people and locality.

In religions that bind people firmly to place the gods appear to have the following characteristics in common. They have no power beyond the vicinity of their particular abodes; they reward and protect their own people but are harmful to strangers; they belong to a hierarchy of beings that extends from the living members of a family, with their graded authority, to ancestors and the spirits of dead heroes. Religions of this local type encourage in their devotees a strong sense of the past, of lineage and continuity in place. 'Ancestor worship lies at the core of the practice. Security is gained through this historical sense of continuity rather than by the light of eternal and timeless values as propounded in transcendental and universal religions.

Rootedness was an ideal of the ancient Greeks and Romans. The French scholar Fustel de Coulanges explored this theme in detail more than a century ago. He stressed the importance of piety and of ancestor worship. A son was obliged to make sacrifices to the souls of the dead, those of his dead father and other ancestors. To fail in this duty was to commit the greatest act of impiety. An ancestor became a protecting god if provisions were carried to his tomb on the appointed days. He was good and provident to his own family but hostile to those who had not descended from him, driving them from his tomb, inflicting diseases upon them if they approached. Love for one's own kin and hostility, rather than mere indifference, to strangers was a common trait of place-bound religions. Each family had its sacred fire which represented the ancestors. A sacred fire "was the providence of a family, and had nothing in common with the fire of a neighboring family, which was another providence."[4] The altar or family hearth symbolized sedentary life. It must be placed on the ground, and once established it could not be moved except as the consequence of unforeseen necessity. Duty and religion required that the family remain grouped around its altar; the family was as much fixed to the soil as the altar itself. The city was a confederation of families. Just as each family had its fixed hearth, so the city had its hearth in the council house, where the officials and a few especially honored citizens took their meals.[5]

The people of ancient Greece and Italy believed in exclusiveness. Space had its inviolable bounds. Every domain was under the eyes of household divinities, and an uncultivated band of soil marked its limit. On certain days of each month and year the father of the family walked around his field. "He drove victims before him, sang hymns, and offered sacrifices. By this ceremony he believed he had awakened the benevolence of his gods towards his field and his house. . . . The path which the victims and prayers had followed were the inviolable limit of the domain."[6]

In antiquity land and religion were so closely associated that a family could not renounce one without yielding the other. Exile was the worst of fates, since it deprived a man not only of his physical means of support but also of his religion and the protection of laws guaranteed by the local gods. In Euripides's play, *Hippolytus*, Theseus would not impose the death penalty on Hippolytus because swift death was regarded as too light a punishment for his heinous crime. Hippolytus had to drain the bitter dregs of his life as an exile on strange soil, this being the proper fate for the impious.[7]

The Greeks valued autochthony. Athenians took great pride in being natives, in the fact that they could trace their long and noble lineage in one locality. Pericles proclaimed, "Our ancestors deserve praise, for they dwelt in the country without break in the succession from generation to generation, and handed it down free to the present time by their valor."[8] Isocrates argued that Athens was great for many reasons but that her strongest title to distinction lay in the people's autochthony and racial purity. He declaimed:

We did not become dwellers in this land by driving others out of it, nor by finding it uninhabited, nor by coming together here a motley horde composed of many races; but we are of a lineage so noble and so pure that throughout our history we have continued in possession of the very land which gave us birth, since we are sprung from its very soil and are able to address our city by the very names which we apply to our nearest kin; for we alone of all the Hellenes have the right to call our city at once nurse and fatherland and mother.[9]

This profound attachment to the homeland appears to be a worldwide phenomenon. It is not limited to any particular culture and economy. It is known to literate and nonliterate peoples, hunter-gatherers, and sedentary farmers, as well as city dwellers. The city or land is viewed as mother, and it nourishes; place is an archive of fond memories and splendid achievements that inspire the present; place is permanent and hence reassuring to man, who sees frailty in himself and chance and flux everywhere.

"The Maori [in New Zealand]," Raymond Firth wrote, "had a great respect for land *per se*, and an exceedingly strong affection for his ancestral soil, a sentiment by no means to be correlated only with its fertility and immediate value to him as a source of food. The lands whereon his forefathers lived, fought, and were buried were ever to him an object of the deepest feeling. . . . 'Mine is the land, the land of my ancestors' was his cry."[10] The Maori revealed their deep-rooted affection in a number of ways. For example, a prisoner, when about to be slain, might ask to be conducted first to the border of his tribal territory so that he could look upon it once again before death. "Or he might ask that he should be allowed to drink of the waters of some stream which flowed through the borders of his home."[11] Tales of heroic deeds added respect to affection for land. Among the most important of these tales were accounts of the arrival of ancestral çanoes in New Zealand more than twenty generations ago.[12]

European students are acquainted with the speeches of Pericles and Isocrates in which these patriots proclaimed their piety for Athens and the Athenians. In the United States, where knowledge of classical antiquity is less emphasized, students may nonetheless acquire a feeling for what profound attachment to ancestral land can mean in the eloquent address of an Indian chief. On the sad occasion when native Americans had to cede land to Governor Stevens of Washington Territory, an Indian chief is reported to have said:

There was a time when our people covered the whole land as the waves of a wind-ruffled sea covers its shell-paved floor, but that time has long since passed away with the greatness of tribes now almost forgotten. I will not dwell on nor mourn over our untimely decay, nor reproach my pale-face brothers with hastening it. We are two distinct races. There is little in common between us. To us the ashes of our ancestors are sacred and their final resting place is hallowed ground, while you wander far from the graves of your ancestors, and, seemingly, without regret. . . . Every part of this country is sacred to my people. Every hillside, every valley, every plain and grove has been hallowed by some fond memory or some sad experience of my tribe. Even the rocks, which seem to lie dumb as they swelter in the sun along the silent seashore in solemn grandeur, thrill with memories of past events connected with the lives of my people. The very dust under your feet responds more lovingly to our footsteps than to yours, because it is the ashes of our ancestors, and our bare feet are conscious of the sympathetic touch, for the soil is rich with the life of our kindred.[13]

Profound sentiment for land has not disappeared; it persists in places isolated from the traffic of civilization. The rhetoric of sentiment barely alters through the ages and differs little from one culture to another. Consider the meaning of the German word *Heimat* as given in a South Tyrolean almanac for the year 1953. Leonard Doob, who discovered this superb specimen of Heimat sentimentality in our time, provides the following translation:

Heimat is first of all the mother earth who has given birth to our folk and race, who is the holy soil, and who gulps down God's clouds, sun, and storms so that together with their own mysterious strength they prepare the bread and wine which rest on our table and give us strength to lead a good life. . . . Heimat is landscape. Heimat is the landscape we have experienced. That means one that has been fought over, menaced, filled with the history of families, towns, and villages. Our Heimat is the Heimat of knights and heroes, of battles and victories, of legends and fairy tales. But more than all this, our Heimat is the land which has become fruitful through the sweat of our ancestors. For this Heimat our ancestors have fought and suffered, for this Heimat our fathers have died.[14]

Rootedness in the soil and the growth of pious feeling toward it seem natural to sedentary agricultural peoples. What of nomadic hunters and gatherers? Because they do not stay in one place and because their sense of land ownership is ill-defined, we might expect less attachment; but in fact the strongest sentiment for the nurturing earth can exist among such people. American Plains Indians have migratory habits. The Comanches, for example, change the location of their principal encampment from year to year, yet they worship the earth as mother. It is for them the receptacle and producer of all that sustains life; in honor it is second only to the sun. Mother earth is implored to make things grow so that they may eat and live, to make the water flow so that they may drink, and to keep the ground firm so that they can walk on it.[15] The Lakota of the Northern Plains have the warmest feeling for their country, particularly the Black Hills. A tribal legend describes these hills as a reclining female from whose breasts issue life-giving forces, and to them the Lakota go like children to their mother's arms. The old people, even more than the young, love the soil; they sit or recline on the ground so as to be close to a nurturing power.[16]

The attitude of American Plains Indians may be influenced by their own agricultural past or by contact with agriculturalists. Australian aborigines, who cannot have been affected by the values of soil tillers, provide a clear example of how hunters and gatherers can be intensely attached to place. Aborigines have no rules of landownership and no strict ideas of territorial boundary. They do, however, distinguish two types of territory—"estate" and "range." Estate is the traditionally recognized home or dreaming place of a patrilineal descent group and its adherents. Range is the tract or orbit over which the group ordinarily hunts and forages. Range is more important than estate for survival; estate is more important than range for social and ceremonial life. As the aborigines put it, range is where they could walk about or run; estate is where they could sit. Strong emotional ties are established with the estate. It is the home of ancestors, the dreaming place where every incident in legend and myth is firmly fixed in some unchanging aspect of nature—rocks, hills and mountains, even trees, for trees can outlive human generations. In times of scarcity, which are frequent along the margins of the desert, the people will leave their own range to forage in other groups' ranges, but seldom for long.[17] As a member of the Ilbalintja tribe explained to the anthropologist Strehlow, "Our fathers taught us to love our own country, and not to lust after the lands belonging to other men. They told us that Ilbalintja was the greatest bandicoot totemic centre amongst the Aranda people, and that, in the beginning, bandicoot ancestors had come from every part of the tribe to Ilbalintja alone and had stayed there for ever: so pleasing was our home to them."[18]

Landscape is personal and tribal history made visible. The native's identity—his place in the total scheme of things—is not in doubt, because the myths that support it are as real as the rocks and waterholes he can see and touch. He finds recorded in his land the ancient story of the lives and deeds of the immortal beings from whom he himself is descended, and whom he reveres. The whole countryside is his family tree.[19]

Modern society has its nomads—hoboes, migrant workers, and merchant seamen, among others. What are the consequences of rootlessness? Do they long for a permanent place, and if so, how is this longing expressed? Migrant workers with their families adapt to the nomadic life out of necessity, not choice. Merchant seamen, in contrast, opt for the sea and rootless wayfaring. They may join the merchant marine in their teens or in early manhood. The ship is their home, the mates are their family, yet there appears to be a craving for a permanent locality as an anchor for their imagination when out at sea. Robert Davis, in an unpublished M.A. thesis, wrote of the seamen he knew personally thus:

They had a craving for a headquarters somewhere along the shore, a place where they could leave their trunk, if they had one; a place to which they could project their minds, wherever they might wander, and visualize the position of the furniture, and imagine just what the inmates of the place were doing at the different hours of the day; a place to which they could send a picture postcard or bring back a curio; a place to which they could always return and be sure of a welcome.[20]

Attachment to the homeland is a common human emotion. Its strength varies among different cultures and historical periods. The more ties there are, the stronger is the emotional bond. In antiquity both the city and the countryside may be sacred, the city because of its shrines, which house local gods and heroes, the countryside because of its nature spirits. But people live in the city and form emotional ties of other kinds, whereas they do not live in the sacred mountains, springs, or groves. Sentiment for nature, inhabited only by spirits, is therefore weaker. A people may, however, become strongly attached to a natural feature because more than one tie yoke them to it. As an example, consider the peak of Reani, the crowning point of the island of Tikopia in the South Pacific. This peak is a landmark of singular importance to the seafaring islanders for at least three reasons. First, it enables the ocean rover to estimate how far he is from land and whether he is on course; this is the practical reason. Second, it is an object of sentiment: the wanderer, when he departs, loses sight of the peak below the ocean waves in sorrow, and, when he returns, greets its first appearance above the waves with joy. Third, it is a sacred place: "it is there that the gods first stand when they come down."[21]

A homeland has its landmarks, which may be features of high visibility and public significance, such as monuments, shrines, a hallowed battlefield or cemetery. These visible signs serve to enhance a people's sense of identity; they encourage awareness of and loyalty to place. But a strong attachment to the homeland can emerge quite apart from any explicit concept of sacredness; it can form without the memory of heroic battles won and lost, and without the bond of fear or of superiority vis-à-vis other people. Attachment of a deep though subconscious sort may come simply with familiarity and ease, with the assurance of nurture and security, with the memory of sounds and smells, of communal activities and homely pleasures accumulated over time. It is difficult to articulate quiet attachments of this type. Neither the rhetoric of an Isocrates nor the effusive prose of a German *Volkskalender* seems appropriate. Contentment is a warm positive feeling, but it is most easily described as incuriosity toward the outside world and as absence of desire for a change of scene. To illustrate this deep undramatic tie to locality, consider three human groups of widely divergent geographical and cultural milieus: the primitive Tasaday of the Mindanao rain forest in the Philippines; the ancient Chinese (their attitude revealed in a Taoist classic); and a modern American farm family in northwestern Illinois.

The outside world discovered the Tasaday in 1971. As yet very little is known about them. They appear to have lived for generations in complete isolation, even from tribes that share the Mindanao rain forest with them. Their material as well as mental culture is perhaps among the simplest in the world. They are food gatherers; their hunting skills are elementary. They seem to lack rituals, ceremonials, or any kind of systematic world view. They are not curious to know about the world beyond the small confines of their homeland. Their language contains no word for sea or lake, although the Celebes Sea and Lake Sebu are less than forty miles away.[22]

> "Why didn't you leave the forest?"
> "We can't go out of our place."
> "Why?"
> "We love to stay in our forest.
> We like it here. It is a quiet place to sleep.
> It is warm. Not loud."[23]

In China the ideal of the simple and sedentary life is stated in the Taoist classic, the *Tao Te Ching*. One passage in it reads: "Let us have a small country with few inhabitants. . . . Let the people return to the use of knotted cords [for keeping records]. Let their food be sweet, their clothing beautiful, their homes comfortable, their rustic tasks pleasurable. The neighboring state might be so near at hand that one could hear the cocks crowing and dogs barking in it. But the people would grow old and die without ever having been there."[24]

The last example is from the American heartland. Six generations of a farm family—the Hammers—have lived and died in Daviess County, northwestern Illinois. Here is a people for whom the riches and wonders of the outside world do not beckon. One middle-aged Hammer explained: "My dad never traveled far and I don't have to. We have so many kinds of recreation right on our own farm. We have a nice stream for fishing, we have hunting. I can hunt deer, squirrels, rabbits— anything you want to hunt. I got them here, right on the farm. I don't have to travel."[25] Young Bill Hammer and Dorothy, married in 1961, went to California for their honeymoon but quickly returned because, as Dorothy put it, "It's so unreal to be gone."[26] Loyalty to the homeland is taught in childhood. In 1972, nine-year-old Jim Hammer was asked what his mother had taught him. He replied:

"What did Mom teach me? For one thing, she taught me how to mow the lawn. She showed me how to tie my shoes. . . . And she tries to teach me to live decent. Like some people don't have a very good life because they don't settle down in one place and don't stay very long. They could live in Illinois for a while and then move to California. I like Illinois; it's just my home state."[27]

12
Visibility: the Creation of Place

Place can be defined in a variety of ways. Among them is this: place is whatever stable object catches our attention. As we look at a panoramic scene our eyes pause at points of interest. Each pause is time enough to create an image of place that looms large momentarily in our view. The pause may be of such short duration and the interest so fleeting that we may not be fully aware of having focused on any particular object; we believe we have simply been looking at the general scene. Nonetheless these pauses have occurred. It is not possible to look at a scene in general; our eyes keep searching for points of rest. We may be deliberately searching for a landmark, or a feature on the horizon may be so prominent that it compels attention. As we gaze and admire a famous mountain peak on the horizon, it looms so large in our consciousness that the picture we take of it with a camera is likely to disappoint us, revealing a midget where we would expect to find a giant.

The peak on the horizon is highly visible. It is a monument, a public place that can be pointed to and recorded. In America the first scenic spots to be appreciated had dramatic presence: the water gap, the gorge, the natural bridge, and—in Yellowstone—geysers. A natural feature may be inconspicuous and yet become a place of sufficient importance to attract

Visibility: the Creation
of Place

tourists. The source of the Mississippi River, for example, is not eye-catching; it is a small body of water like the thousands of lakes and springs in the same region. Only scientists, after detailed measurements, can tell which pool is the source. Once a particular body of water was marked as the Mississippi's source and the area around it designated a park, it became a place to which people would want to visit and have their pictures taken. Scientists thus appear to have a certain power: they can create a place by pointing their official fingers at one body of water rather than another.

Many places, profoundly significant to particular individuals and groups, have little visual prominence. They are known viscerally, as it were, and not through the discerning eye or mind. A function of literary art is to give visibility to intimate experiences, including those of place. The Grand Tetons of landscape do not require the services of literature; they advertise themselves by sheer size. Literary art can illuminate the inconspicuous fields of human care such as a Midwestern town, a Mississippi county, a big-city neighborhood, or an Appalachian hollow.

Literary art draws attention to areas of experience that we may otherwise fail to notice. Sculptures have the power to create a sense of place by their own physical presence (Fig. 18). A single inanimate object, useless in itself, can be the focus of a world. Wallace Stevens wrote in a poem that a jar placed on a hill "made the slovenly wilderness surround that hill." The jar took dominion. "The wilderness rose up to it, and sprawled around, no longer wild."[1] The human being can command a world because he has feelings and intentions. The art object may seem to do so because its form, as Langer would say, is symbolic of human feeling.[2] A piece of sculpture appears to incarnate personhood and be the center of its own world. Although a statue is an object in our perceptual field, it seems to create its own space.

We can refuse significance to the jar on the hill; the jar simply occupies space and does not command it. Objects that are held in awe by one people can easily be overlooked by another. Culture affects perception. Yet certain objects, both

Visibility: the Creation
of Place

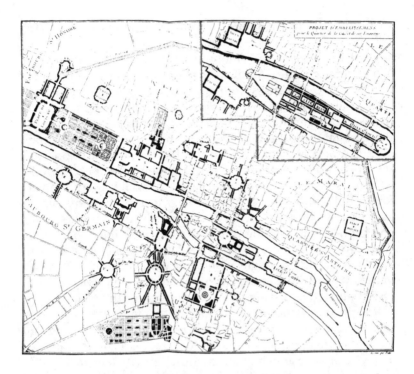

Figure 18. Place as highly visible public symbol, a feature that architects can create. M. Patte's prize-winning plan for the Paris of Louis XV, in which the *place royale* is of great prominence. Each *place royale* has a statue of the monarch at the center, and streets fanning out like rays.

natural and man-made, persist as places through eons of time, outliving the patronage of particular cultures. Perhaps any large feature in the landscape creates its own world, which may expand or contract with the passing concerns of the people, but which does not completely lose its identity. Ayers Rock in the heart of Australia, for example, dominated the mythical and perceptual field of the aborigines, but it remains a place for modern Australians who are drawn to visit the monolith by its awesome bulk (Fig. 19A). Stonehenge is an architectural example. No doubt it is less a place for British tourists than for its original builders: time has caused its dread as well as its stones to erode, but Stonehenge remains very much a place (Fig. 19B).[3]

How is it possible for a monument to transcend the values of a particular culture? An answer might be: a large monument like Stonehenge carries both general and specific import. The specific import changes in time whereas the general one remains. Consider the modern Gateway Arch of St. Louis. It has the general import of "heavenly dome" and "gate" that transcends American history, but it also has the specific import of a unique period in American history, namely, the opening of the West to settlement. Enduring places, of which there are very few in the world, speak to humanity. Most monuments cannot survive the decay of their cultural matrix. The more specific and representational the object the less it is likely to survive: since the end of British imperialism in Egypt, the statues of Queen Victoria no longer command worlds but merely stand in the way of traffic. In the course of time, most public symbols lose their status as places and merely clutter up space.

If a piece of sculpture is an image of feeling, then a successful building is an entire functional realm made visible and tangible. As Langer put it, "The architect creates a culture's image: a physically present human environment that expresses the characteristic rhythmic functional patterns which constitute a culture."[4] The patterns are the movements of personal and social life. They are fluid and enormously complex. It is hardly possible to specify them in detail and design accordingly. An architect has an intuitive grasp, a tacit understanding, of the rhythms of a culture, and he seeks to give them symbolic form. A house is a relatively simple building. It is a place, however, for many reasons. It provides shelter; its hierarchy of spaces answers social needs; it is a field of care, a repository of memories and dreams. Successful architecture "creates the semblance of that World which is the counterpart of a Self."[5] For personal selfhood that world is the house; for collective selfhood it is a public environment such as temple, town hall, or civic center.

Art and architecture seek visibility. They are attempts to give sensible form to the moods, feelings, and rhythms of func-

Visibility: the Creation of Place

ENDURING PLACES — MONUMENTS

A. Ayers Rock

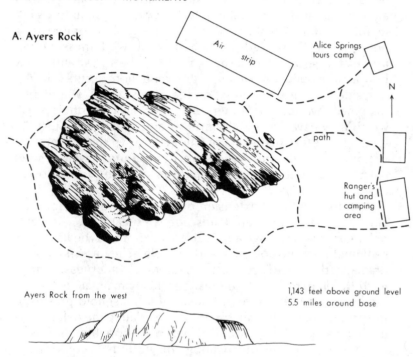

Ayers Rock from the west

1,143 feet above ground level
5.5 miles around base

B. Trilithons of Stonehenge from the east

Height: 16.5 - 22 feet

Figure 19. Enduring places: Ayers Rock in the heart of Australia and Stonehenge at the central node of southern England.

Visibility: the Creation
of Place

tional life. Most places are not such deliberate creations. They are built to satisfy practical needs. How do they acquire visibility for both local inhabitants and outsiders? Think of the way a new country is settled. At first there is wilderness, undifferentiated space. A clearing is made in the forest and a few houses are built. Immediately differentiation occurs; on the one side there is wilderness, on the other a small, vulnerable, man-made world. The farmers are keenly aware of their place, which they have created themselves and which they must defend against the incursions of wild nature. To the passerby or visitor, the fields and houses also constitute a well-defined place, obvious to him as he emerges from the forest to the clearing.

With the continual extension of clearings the forest eventually disappears. An entire landscape is humanized. The fields belonging to one village adjoin those of another. The limits of a settlement are no longer clearly visible. They are no longer dramatized by the discernible edges of the wilderness. Henceforth the integrity of place must be ritually maintained. In the time of republican Rome the head of a household preserved the borders of his domain by circumambulating the fields, singing hymns, and driving sacrificial victims before him. In Britain the ancient custom of "beating the bounds" required the parish priest to walk around the parish and strike certain markers with a stick. In the Netherlands the village of Anderen is a deeply rooted community. As late as 1949 village elders and teenaged youths continued the annual practice of inspecting the boundary markers. The elders, to ensure that the young would not forget the exact location of the markers, boxed the youngsters' ears.[6]

To the casual visitor the limits of village domain are not evident in the landscape. The villages themselves are evident, each surrounded by an apron of fields. To the local people sense of place is promoted not only by their settlement's physical circumscription in space; an awareness of other settlements and rivalry with them significantly enhance the feeling of uniqueness and of identity. French villages, such as those in Lorraine, Burgundy, Champagne, and Picardy, are nucleated

Visibility: the Creation
of Place

settlements with (often) a church at the center. Peasants
socialize in the winter evenings and again on holidays and
market days. They work together at harvest time and during the
vintage. The casual observer may conclude that the village is
one place, a unified community conscious of its identity vis-à-
vis neighboring communities. This is true, yet the village itself
is divided. Egoism and contentious pride exist within each settlement as well as between settlements. Maurice Halbwachs
notes: "Just as a village sometimes ignores, envies and detests
a neighboring village, so it happens only too often that families
envy each other from one house to the next, without ever a
thought of helping each other. . . . There is no natural tendency to work together for the common good."[7] Egoism and
envy are reprehensible traits. However, they promote a conscious sense of self and of the things associated with self, including home and locality.

The question of how the awareness of place in a rural region
varies with scale is clarified in William Skinner's work on traditional China (Fig. 20). Skinner believes that "insofar as the
Chinese peasant can be said to live in a self-contained world,
that world is not the village but the standard marketing community."[8] The area of a standard marketing community is
about twenty square miles. Within it live some seven to eight
thousand people distributed among a score or so of settlements. The typical peasant sees his fellow villagers far more
often than he does outsiders; his own village is his primary
place. Nonetheless a peasant, by the time he is forty to fifty
years old, has visited the local market town several thousand
times, and in its teahouses he has socialized with peasants
from village communities far removed from his own. A
middle-aged villager has a nodding acquaintance with almost
every adult in all parts of the marketing system.[9] He is aware,
then, of a social world much larger than his own village community. Does he also know this larger world as a bounded
region, a place with distinctive traits that set it apart from other
comparable units?

The boundary of a nucleated settlement is clearly visible. In
contrast, the outer edge of the standard marketing system is

*Visibility: the Creation
of Place*

AWARENESS OF PLACE AT DIFFERENT SCALES

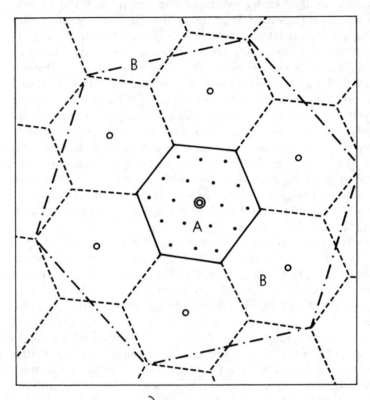

- village
  } visibly bounded places
○ market town

A marketing area; distinctive place but without
  tangible boundary

B marketing region; conceptual "place"

Figure 20. Awareness of place at different scales. The villages and market towns
are visibly bounded places in distinction to the marketing area and the marketing region which have no tangible boundaries. Recognizing the marketing area
or region as "place" is more than a matter of looking as a tourist might look.

not a physical feature that strikes the eye. In traditional rural China the marketing area is often a close-knit functional unit. Its high degree of self-sufficiency is suggested by the fact that weights and measures, and even language, show perceptible differences from those of adjacent marketing communities. But do the local people know this fact? The local elite are probably aware of it. Landlords visit not only the neighborhood town but also the town of a higher order in size and function where their special needs, such as books, can be satisfied. From the perspective of the higher-order place, the elite may well discern that their own marketing area is one among several. The standard marketing area is integrated by many activities. Only one, however, is highly visible. This is the religious procession that defines the earthly domain of the temple god. The procession has the effect of dramatizing the marketing area as bounded space. It is "an annual reaffirmation of the community's territorial extent and a symbolic reinforcement of its town-centered structure."[10]

The clustered village stands out in the landscape. Approaching a rural settlement we can see the silhouette of houses and trees rising above the cultivated fields. In comparison, urban neighborhoods lack visual prominence. Each neighborhood is a small part of a large built-up area, and it is unclear where one unit ends and another begins. A planner looking at the city may discern areas of distinctive physical and socioeconomic character; he calls them districts or neighborhoods and assigns them names if local ones do not already exist. These neighborhoods are places for him, they have meaning for him as intellectual concepts. What would be the perception of the people who live in such areas? Will they also see that in their area the houses are of a similar build and that the people are mostly of a similar socioeconomic class? The answer is, of course, not necessarily. Local inhabitants have no reason to entertain concepts that are remote from their immediate needs.

The lack of a concept "neighborhood" corresponding to that of the city planner is well illustrated in Herbert Gans's study of Boston's West End. This old working-class district was declared a slum and torn down under a federal renewal program be-

Visibility: the Creation
of Place

tween 1958 and 1960. Defenders of the district had difficulty marshaling the support of the local people. West Enders never used the term "neighborhood." They showed little concern for the district as a physical and social entity; their interest was confined essentially to their own street and to the stores they frequented.[11] Politicians, recognizing this extreme localism, promised improvements for individual streets rather than for the district as a whole. They did not try to raise the consciousness of their constituents beyond the small world of immediate experience. When the West End as a whole was threatened with demolition, the people were shocked into awareness. Even then some felt sure that while the entire district was coming down, their own street would be spared. The local people who participated in the Save the West End Committee were a handful of intellectuals and artists. Unlike their neighbors, these people did have a concept of "neighborhood." Gans explained: "Although they were active within their own peer groups, their career and creative interests separated them from these groups psychologically. . . . As a result, they developed a strong symbolic identification with the West End. Partially because of their skills and their marginality, they were able to develop a holistic concept of the West End as a neighborhood."[12]

The street where one lives is part of one's intimate experience. The larger unit, neighborhood, is a concept. The sentiment one has for the local street corner does not automatically expand in the course of time to cover the entire neighborhood. Concept depends on experience, but it is not an inevitable consequence of experience. The concept can be elicited and clarified by questioning, directed first at the concrete and then at the more abstract. Questions and answers may proceed in the following manner:

What is or what constitutes *my* neighborhood?
Answer: It is where I live and where I go shopping; from which I gather that each person has his own neighborhood.
What is *our* neighborhood?
Answer: It is the locale of my own kind of people, that is to say, the Irish in a mixed Italian-Irish working-class area.

Visibility: the Creation
of Place

What is *the* neighborhood?
Answer: It is the Italian-Irish working-class area,
a physical and social unit that I am vaguely aware
of as different from adjoining areas.

The larger unit acquires visibility through an effort of the mind. The entire neighborhood then becomes a place. It is, however, a conceptual place and does not involve the emotions. Emotion begins to tinge the whole neighborhood—drawing on, and extrapolating from, the direct experience of its particular parts—when the neighborhood is perceived to have rivals and to be threatened in some way, real or imagined. Then the warm sentiment one has for a street corner broadens to include the larger area. Although an external event, such as urban renewal, enables a people to see the larger unit, this perception becomes vividly real if the unit, in fact, has strong local flavor, visual character, and clear boundaries. Houses and streets do not of themselves create a sense of place, but if they are distinctive this perceptual quality would greatly help the inhabitants to develop the larger place consciousness.

Working-class and poor people do not live in homes and neighborhoods of their own design. They move either into residences that have been abandoned by the well-to-do, or into new subsidized housing. In both cases the physical structures do not reflect their dwellers' ideals. Sentiment, if it exists, has developed as slowly as familiarity. In contrast, the affluent are able to occupy an environment of their own design. Their dreams are quickly convertible into houses and lawns. From the start the affluent can live in a place of their own, surrounded by their own kind of people, and they are well aware of this fact. The rich neighborhood is, from the start, highly visible to both residents and outsiders. Its architecture is likely to show character and the grounds may be walled off, with a guard at the gate.

Beacon Hill in Boston is a famous old neighborhood. It began, however, as a suburban dream of affluent Bostonians living in the post-Independence decades. Now steeped in history, it was once a showy residential estate. The rich sense of place at Beacon Hill and its high visibility result from a combi-

Visibility: the Creation
of Place

nation of factors. Architectural distinction is one; houses are of a style that differentiates them from buildings in adjoining areas. Time is another; time has given Beacon Hill residents long memories. Notable events and persons are a third; they have given the neighborhood luster. Kin and neighborhood ties are strong, expressed not (of course) in the borrowing of cups of sugar but in social calls and the exchange of intimate dinners. Residents are proud of the place's traditions; they have the leisure and the education to produce a pamphlet literature that tastefully draws people's attention to the neighborhood's heritage. Public rites enhance Beacon Hill's visibility. At Christmastime, for example, the display of candle lights in the homes attracts a large number of tourists. These informal means of promoting the place's identity are supplemented by the effort of formal organizations such as the Beacon Hill Association, which was founded for the purpose of keeping undesirable people and enterprise out of the area.[13]

A district's reputation may depend far more on the propaganda of outside groups than of local residents. Even Greenwich Village, rich in artists whose calling is to articulate values, owes its Bohemian image not a little to promotion by outside media and real estate agents.[14] Slums and skid rows are distinctive places in many large North American cities. Some are so peculiar from the standpoint of middle-class values that they become tourist attractions. Air-conditioned buses take upright small-town citizens through Chicago's skid row as if it were a titillating peep show. Derogatory names like "Jew Town," "Nigger Town," and "Back of the Yard" are imposed by fearful outsiders on the local inhabitants. At first the local people may not themselves be aware of their membership in the larger neighborhood; they know only that they live on a certain block in the poorer part of the city. In time, however, the outside message sinks in. The local people begin to see that they live in, say, "Back of the Yard," an area with a certain character and with boundaries that outsiders fear to cross. "Back of the Yard" as a whole becomes a shadowy reality for the residents, a reality viewed with a mixture of helplessness, resentment, and

Visibility: the Creation
of Place

perhaps also pride if the possibility for political action goes with the consciousness of place.

The city is a place, a center of meaning, par excellence. It has many highly visible symbols. More important, the city itself is a symbol. The traditional city symbolized, first, transcendental and man-made order as against the chaotic forces of terrestrial and infernal nature. Second, it stood for an ideal human community: "What is the Citie, but the People? True, the People are the Citie" (Shakespeare, *Coriolanus*, act 3, scene 1). It was as transcendental order that ancient cities acquired their monumental aspect. Massive walls and portals demarcated sacred space. Fortifications defended a people against not only human enemies but also demons and the souls of the dead. In medieval Europe priests consecrated city walls so that they could ward off the devil, sickness, and death—in other words, the threats of chaos.[15]

A city draws attention to itself, achieving power and eminence through the scale and solemnity of its rites and festivals. Ancient capitals began as ritual centers of high import. Splendid architectural settings were required for the enactment of sacred dramas. In time ceremonial centers attracted secular population and activities. Economic functions multiplied and submerged the city's religious identity. However, the feeling for drama and display remained as did the form and style of religious rites which branched into the secular sphere. In medieval Europe the cathedrals and churches, far more vividly colored then than now, were the centers of celebrations that punctuated the church's calendar year. Secular events called for display no less than religious ones. In medieval London, crowds turned out not only on royal occasions, but also for visitations by far lesser dignitaries; even the progress of a prisoner to jail was cause for a festive mood in the streets.[16]

It hardly needs saying that the visibility of a modern city suffers from the lack of public occasions to which the people are drawn and for which the halls and streets function as supportive stage. Of course the city was and is an elaborate conglomeration of innumerable stages for the performance of pri-

Visibility: the Creation
of Place

vate and semi-public dramas—birthdays, high school gradua-
tions, basketball tournaments—but these are at most local
pageantries often held at some distance from the city core.
Ceremonials such as laying the cornerstone of a civic building,
planting a tree in the public square, and consecrating a church
seem to have become increasingly empty gestures of another
age, to which the busy and skeptical citizens of today attach
little meaning. In the nineteenth and early decades of the
twentieth century urban Americans still had a sense of occasion, a feeling that certain city events called for some form of
public festivity. Consider Minneapolis. In 1896 Colonel Stevens's house, the first house built within the city limits, was put
on wheels and pulled by relay teams of nearly ten thousand
school children from its original site near Hennepin Bridge to
Minnehaha Park. It was an occasion that stirred the local
citizenry. People lined the streets to watch the house go by.
Such an event will hardly excite sophisticated Minneapolitans
today. Here is another illustration. When the city's Foshay
Tower was completed in 1929 its owner saw fit to invite the
governors of the forty-eight states to attend its opening ceremony. On the other hand, when the IDS building was completed in 1972 it became Minneapolis's tallest skyscraper and
preeminent landmark, yet its opening passed with little fan-
fare.

A city does not become historic merely because it has occupied the same site for a long time. Past events make no
impact on the present unless they are memorialized in history
books, monuments, pageants, and solemn and jovial festivities
that are recognized to be part of an ongoing tradition. An old
city has a rich store of facts on which successive generations of
citizens can draw to sustain and re-create their image of place.
Confident of their past, citizens can afford to speak with a soft
voice and go about the business of putting their hometown on
a pedestal with taste. New cities, such as the frontier settlements of North America, lacked a venerable past; to attract
business and gain pride their civic leaders were obliged to
speak with a loud voice.[17] Strident boosterism was the tech-

Visibility: the Creation
of Place

nique to create an impressive image, and to a lesser extent it still is. The boosters could rarely vaunt their city's past or culture; hence the emphasis tended to be on abstract and geometrical excellences such as "the most central," "the biggest," "the fastest," and "the tallest." Boosterism has by now become something of an American tradition, and it is practised with the panache of a Pop-art form. Jan Morris, in an article on Tennessee, asked:

Did you know that Chattanooga had the biggest Sunday School in the world? that it is the Electrical Capital of the world? that it supports more churches per head than anywhere else in the world? that the steepest funicular railway in the world runs up Lookout Mountain to the highest railway station in America? that the Choo-Choo Restaurant is the biggest eating-house in the world? that Chattanooga is America's Saddlery? that the view you are enjoying is the Longest View in the South, embracing seven States? "Made in Chattanooga," says the proud boast in many a local store, "by Chattanoogans"—which is to say, created on the spot by the Brightest and Best of the Sons of the Morning.[18]

Sense of self, whether individual or collective, grows out of the exercise of power. Cities may have achieved their maximum visibility as independent political units, that is, as city-states. Take Greek city-states for example. Several factors contributed to their vivid personalities. One was small size. Even Attica, dominated by Athens, was small enough so that its most distant parts could be reached in two long days' walk. Sparta grew ungainly through conquests, but most states were smaller than Sparta.[19] A Greek polis was not an abstract entity: a citizen could know it personally. Even if he had not paced the country from end to end, he should at least be able to see the physical limits of the state to which he owed allegiance. In the clear air he might discern the chain of hills beyond which lay other states that competed with his own. Another factor that enhanced the city's sense of self was the small size of its population. People learned to know one another. A wide net of social communication does not in itself generate public enterprise. The Greeks, however, believed that manhood demanded full participation in the functions of the state, whether

Visibility: the Creation
of Place

as administrators or as soldiers. Public service and the winning of glory far outranked the quiet and often traceless satisfactions of private life.

Competition among the city-states fueled patriotic fervor and promoted in each state a heightened awareness of its own individuality. Competition took the form of wars and athletic contests. Wars were fought for territory and for dominion over a weaker neighbor. Athletic rivalry flared every four years at Olympia in honor of Zeus. The games were extremely nationalistic in spirit; cities were more vain of victories won at Olympia than on battle fields. Quieter ways to boast also existed. Athens was proud of its government. As Pericles put it, "Our form of government does not enter into rivalry with the institution of others. We do not copy our neighbors, but are an example to them."[20] Sparta was proud of its citizen soldiers; unlike other cities it did not need a physical wall for defense.

Tyrants promoted the identity of their capital cities. In ancient Greece tyrannies emerged in response to a pressing need to restore order to the state. The tyrant, to keep his position, must seek public approval. He had two tried methods of winning it. One was adventure abroad; foreign wars fostered national sentiment and at the same time made the people forget their political servitude. The other method called for munificence, such as large-scale public works, including temple construction, and the subsidization of art. Grand artworks provided a focus and an outlet for patriotic zeal.[21]

The city-state was small enough that most of its citizens could know it personally. The modern nation-state is far too large to be thus experienced. Symbolic means had to be used to make the large nation-state seem a concrete place—not just a political idea—toward which a people could feel deep attachment. The belief that the nation demands the supreme loyalty of man is a modern passion. Since the end of the eighteenth century it has infected more and more people throughout the world. Despite universalist ideals on the one hand and the pull of localism on the other, the nation-state is now the world's dominant political unit. To be a modern nation, local

Visibility: the Creation
of Place

attachments based on direct experience and intimate knowledge have to be overcome. Thus Ernst Moritz Arndt (1769–1860), an early apostle of German nationalism, wrote:

> Where is the German's Fatherland?
> Is it Swabia? Is it the Prussian land?
> Is it where the grape grows on the Rhine?
> Where sea-gulls skim the Baltic's brine?
> O no! more great, more grand
> Must be the German's Fatherland![22]

The sentiment that once tied people to their village, city, or region had to be transferred to the larger political unit. The nation-state, rather than any of its parts, was to achieve maximum visibility. How could this be done? One method was and is to make the state the object of a religious cult. The French Legislative Assembly decreed in June 1792 that "in all the communes an altar to the Fatherland shall be raised, on which shall be engraved the Declaration of Rights with the inscription, 'the citizen is born, lives, and dies for *la Patrie*.'"[23] In patriotic fervor men say, "We must protect our sacred soil." They are saying, in effect, that "the land which is our country must be protected as if all of it were like a church." The field and the cesspool upon the land are details, mundane and irrelevant.[24] To make the idea of the sacred country seem real, sacred places that can be directly experienced are created. In the United States these are not churches and cathedrals. They are places like Independence Hall in Philadelphia, the shrines of General Lee in Lexington and of General Grant in New York, and the stately monuments of the city of Washington.[25]

History books helped to transform the nation-state into place—and indeed into person. Patriotic literature is replete with personifications such as "the national will," and "national destiny." Image-building through history books flourished in the nineteenth century. In earlier periods, as Carleton Hayes observed, history had been local, "world" or religious history, "chronicles of kings, biographies of warriors or saints, philosophical disquisitions upon the course of God's dealings

Visibility: the Creation
of Place

with man, but almost never national history as such. During the nineteenth century, however, very little history was written which was not national in scope or import."[26]

Maps in school atlases and history books show nation-states as sharply bounded units. Small-scale maps encourage people to think of their countries as self-sufficient, discrete entities. Visible limits to a nation's sovereignty, such as a row of hills or a stretch of river, support the sense of the nation as place. From the air, however, mountains and rivers are merely elements of physical geography, and man-made markers like fences and guard posts are invisible. Aerial photographs are useless in history books. Maps, which also present the vertical view, are another matter. Cartography can clearly be made to serve a political end. In a school atlas the world's nations appear as a mosaic of clashing colors. Pink Canada looms large over butter-tinted United States; there can be no doubt about where one ends and another begins, nor of their sharply contrasting identities.

In summary, we may say that deeply-loved places are not necessarily visible, either to ourselves or to others. Places can be made visible by a number of means: rivalry or conflict with other places, visual prominence, and the evocative power of art, architecture, ceremonials and rites. Human places become vividly real through dramatization. Identity of place is achieved by dramatizing the aspirations, needs, and functional rhythms of personal and group life.

13

Time and Place

How time and place are related is an intricate problem that invites different approaches. We shall explore three of them here. They are: time as motion or flow and place as a pause in the temporal current; attachment to place as a function of time, captured in the phrase, "it takes time to know a place"; and place as time made visible, or place as memorial to times past.

Place is an organized world of meaning. It is essentially a static concept. If we see the world as process, constantly changing, we should not be able to develop any sense of place. Movement in space can be in one direction or circular, implying repetition. A common symbol for time is the arrow; others are the circular orbit and the swinging pendulum. Thus images of space and time merge. The arrow represents directional time but also movement in space to a goal. Goal is both a point in time and a point in space. My goal, let us say, is to be a vice-president in a motorcar company. The goal lies in my future; it is the ultimate place in society I wish to attain. The vice-presidency dominates my hope so that intermediate positions, such as foreman and manager, are mere steps up the ladder (Fig. 21A). I do not expect to remain a foreman for long, hence I hesitate to acquire the accouterments of the job. This type of thinking, which is oriented to a future and compelling goal, may be a characteristic trait in the attitude of a whole people. Consider the Israelites and their view of time. The destination of the Chosen People was the Kingdom of God. All intermediate kingdoms were suspect. Unlike the ancient Greeks the Israelites hesitated to establish a political organization that suggested permanence. Earthly places were all temporary, at best stages on the way to the ultimate goal. Religions of transcendental hope tend to discourage the establishment of place. The message is, don't hang on to what you have; live in the present as if it were a camp or wayside station to the future.[1]

The manager's office may be only two doors from the vice-president's office, but it will take the manager years of hard work to get there. The vice-president's office is a temporal goal. Goal is also a place in space, the promised land on the other side of the ocean or mountain. Months may lapse before the emigrants will reach their destination; however, what seems daunting to them at the start of the trip is not the time but the space that has yet to be traversed. Goal is one of the three categories of place that can be distinguished when movement is in one direction, with no thought of return; the other two are home and camps or wayside stations. Home is the stable world to be transcended, goal is the stable world to be attained, and camps are the rest stops for the journey from one world to the other. The arrow is the appropriate image (Fig. 21A).

Most movements are not major undertakings structured around the antipodal points of home base and goal. Most movements complete a more or less circular path, or swing back and forth like a pendulum (Fig. 21B). In the home pieces

Figure 21. Movement, time, and place: A. Linear paths and places; B. Cyclical/pendulumlike paths and places. A comment on Bii: In ancient China, people probably lived in the city through the winter months. As spring approached they moved out, lived in huts built in the countryside and cultivated land that was divided into a rectangular pattern. After harvest the people returned to the city and were engaged in various services, trades, and crafts. Thus life was divided into two poles—city and countryside, winter and summer, *yin* and *yang*.

A. Linear paths and places

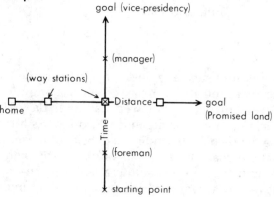

goal (vice-presidency)

✗ (manager)

(way stations)

home ☐——☐——⊠–Distance–☐——→ goal (Promised land)

Time

✗ (foreman)

✗ starting point

B. Cyclical/pendulumlike paths and places

i. Daily

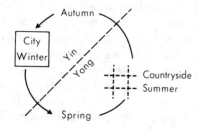

field ← farmstead

game → plant food → camp ← waterhole

office ← cocktail lounge → suburban home

ii. Seasonal (the two poles—places—of ancient China)

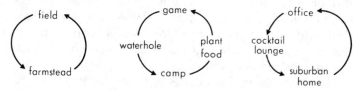

Autumn

City Winter

Yin / Yang

Countryside
Summer

Spring

iii. Stages (places) of life: cyclical model

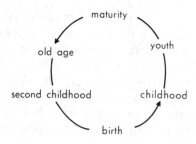

maturity

old age   youth

second childhood   childhood

birth

of furniture such as a desk, an armchair, the kitchen sink, and the swing on the porch are points along a complex path of movement that is followed day after day. These points are places, centers for organizing worlds. As a result of habitual use the path itself acquires a density of meaning and a stability that are characteristic traits of place. The path and the pauses along it together constitute a larger place—the home. While we readily accept our home as a place, we must make an extra effort to realize that smaller places exist within the home. Our attention focuses on the house because it is a clearly circumscribed and visually prominent structure. Walls and the roof give it a unified shape. Remove the walls and the roof and it immediately becomes apparent that such local stations as desk and kitchen sink are themselves important places connected by an intricate path, pauses in movement, markers in routine and circular time.

The nomad's world consists of places connected by a path. Do nomads, who are frequently on the move, have a strong sense of place? Quite possibly. Nomads move, but they move within a circumscribed area, and the distance between the two extreme points of their peregrination seldom exceeds 200 miles.[2] Nomads pause and establish camp at roughly the same places (pastures and water holes) year after year; the paths they follow also show little change. For nomads the cyclical exigencies of life yield a sense of place at two scales: the camps and the far larger territory within which they move. It may be that the camps are their primary places, known through intimate experience, whereas the territory traversed by nomads seems more shadowy to them because it lacks a tangible structure.

In modern society the relation between mobility and a sense of place can be very complicated. Most people achieve a fairly stable position in society by the time they are thirty to forty years old. They settle into a routine of home, office or factory, and holiday resort. These are distinctive places. There is no confusing the rather tedious work in the office with watching television at home; and the two-week holiday by the sea is a long-contemplated event. As the years pass the places of personal significance remain the same; the family goes to Brighton each summer. In time the sense of place extends beyond individual localities to a region defined by these localities. The region, subtended by home, office, and resort, becomes itself a place although it lacks a visible boundary.

Now consider a high-salaried executive. He moves about so much that places for him tend to lose their special character. What are his significant places? The home is in the suburb. He lives there, but home is not wholly divorced from work. It is occasionally a showplace for the lavish entertainment of colleagues and business associates. It is also a workplace, for the busy executive brings his work home. It is not quite a home for the family because the children are boarders at school. The executive has a cottage in the country. The cottage is a place for the entire family, but only briefly during the summer and not every year; it is a "play" home in which nothing very serious happens. The office is a workplace but it is also the executive's home—to the extent that it is the center of his life; he may, moreover, have an apartment in the office building or downtown where he can spend the night. The executive takes periodic trips abroad, combining business with pleasure. He stays at the same hotel, or with the same friends, in Milan, and again in Barbados. The circuits of movement are complex; even then they represent only a stage in the executive's upwardly mobile career. His goal may still lie ahead. His pattern of movement may yet expand and his constellation of places increase before they shrink inevitably with retirement and the onset of age.[3]

The second theme, closely related to the first, is "how long does it take to know a place?" Modern man is so mobile that he has not the time to establish roots; his experience and appreciation of place is superficial. This is the conventional wisdom. Abstract knowledge *about* a place can be acquired in short order if one is diligent. The visual quality of an environment is quickly tallied if one has the artist's eye. But the "feel" of a place takes longer to acquire. It is made up of experiences, mostly fleeting and undramatic, repeated day after day and over the span of years. It is a unique blend of sights, sounds, and smells, a unique harmony of natural and artificial rhythms such as times of sunrise and sunset, of work and play. The feel of a place is registered in one's muscles and bones. A sailor has a recognizable style of walking because his posture is adapted to the plunging deck of a boat in high sea. Likewise, though less visibly, a peasant who lives in a mountain village may develop a different set of muscles and perhaps a slightly different manner of walking from a plainsman who has never climbed. Knowing a place, in the above senses, clearly takes time. It is a subconscious kind of knowing. In time we become familiar with a place, which means that we can take more and more of it for granted. In time a new house ceases to make little demands on our attention; it is as comfortable and unobtrusive as an old pair of slippers.

Attachment, whether to a person or to a locality, is seldom acquired in passing. Yet the philosopher James K. Feibleman noted: "The importance of events in any life is more directly proportionate to their intensity than to their extensity. It may take a man a year to travel around the world—and leave absolutely no impression on him. Then again it may take him only a second to see the face of a woman—and change his entire future."[4] A man can fall in love at first sight with a place as with a woman. The first glimpse of the desert through a mountain pass or the first plunge into forested wilderness can call forth not only joy but, inexplicably, a sense of recognition as of a pristine and primordial world one has always known. A brief but intense experience is capable of nullifying the past so that we are ready to abandon home for the promised land. Still more curious is the fact that people can develop a passion for a certain type of environment without the benefit of direct encounter. A story, a descriptive passage or picture in a book suffices. For example, the scholar C. S. Lewis was overcome with a longing for remoteness and severity, for pure "northernness," when he was a child. Helen Gardner, in an appraisal of Lewis's life and work, wrote:

Northernness [was] a vision of huge, clear spaces hanging above the Atlantic in the endless twilight of Northern summer. Lewis never cooled to his early love. The sadness and sternness of the northern world appealed to something very deep in his nature. But he had never lived in northern lands, nor did he feel the urge to travel north-ward and confront his personal vision with sensuous experience. He had become infatuated with a landscape through literary and musical means such as illustrations to stories from Wagner and records of "The Ring."[5]

Many years in one place may leave few memory traces that we can or would wish to recall; an intense experience of short duration, on the other hand, can alter our lives. This is a fact to bear in mind. Another is this. In relating the passage of time to the experience of place it is obviously necessary to take the human life cycle into account: ten years in childhood are not the same as ten years in youth or manhood. The child knows the world more sensuously than does the adult. This is one reason why the adult cannot go home again. This is also one reason why a native citizen knows his country in a way that cannot be duplicated by a naturalized citizen who has grown up elsewhere. Experienced spans of time, at different stages in life, are not commensurable. The West Indian writer, V. S. Naipaul, makes a character in a novel say this of emigrants:

They went. But they came back. You know, you are born in a place and you grow up there. You get to know the trees and plants. You will never know any other trees and plants like that. You grow up watching a guava tree, say. You know that browny-green bark peeling like old paint. You try to climb that tree. You know after you climb it a few times the bark gets smooth-smooth and so slippery you can't get a grip on it. You get that ticklish feeling in your foot. Nobody has to teach you what the guava is. You go away. You ask, "What is that tree?" Somebody will tell you, "An elm." You see another tree. Somebody will tell you, "That's an oak." Good; you know them. But it isn't the same. Here you wait for the poui to flower one week in the year and you don't even know you are waiting. All right, you go away. But you will come back. Where you born, man, you born.[6]

A young child's experience of time differs from that of an adult. To the young child time does not "flow"; he stands as it were outside it, remaining at the same tender age seemingly forever. To the grown person time rushes on, propelling him forward willy-nilly. Since small children are seldom able to re-flect on their experiences and describe them, we need to make use of the recall and observations of adults. Here is how the playwright Eugene Ionesco recalls his childhood. At the age of eight or nine, everything for him was joy and presentness. Time seemed a rhythm in space. The seasons did not mark the passage of the year; rather they spread out in space. As a young child he stood at the center of a world that was a decorative background, with its colors, now dark, now bright, with its flowers and grass appearing, then disappearing, moving toward him, moving away from him, unfolding before his eyes while he himself stayed in the same place, outside time, watching time pass. At fifteen or sixteen it was all over. The teenaged Ionesco felt as though a centrifugal force had thrown him out of his immutability into the midst of things that come and go and go away for good. He was in time, in flight, in finiteness; the present had disappeared. There was nothing left for him but a past and a tomorrow, a tomorrow that he was already conscious of as past.[7]

Sense of time affects sense of place. To the extent that a small child's time is not that of an older person, neither is his experience of place. An adult cannot know a place as a child knows it, and this is not only because their respective sensory and mental capacities differ but also because their feelings for time have little in common.

As one lives on, the past lengthens. What is this personal past like? Simone de Beauvoir examined her own past and wrote rather gloomily,

The past is not a peaceful landscape lying there behind me, a country in which I can stroll wherever I please, and which will gradually show me all its secret hills and dales. As I was moving forward, so it was crumbling. Most of the wreckage that can still be seen is colorless, distorted, frozen. . . . Here and there, I see occasional pieces whose melancholy beauty enchants me.[8]

What can the past mean to us? People look back for various reasons, but shared by all is the need to acquire a sense of self and of identity. I am more than what the thin present defines. I am more than someone who at this moment is struggling to put thought into words: I am also a published writer, and here is the book, hardbound, resting reassuringly by my side. We are what we have. We have friends, relatives, and ancestors; we have skills and knowledge, and we have done good deeds. But these possessions may be neither visible nor readily accessible. Friends live far away, or have died. Skills and knowledge are not at this time called into use, and may well have rusted. As for worthy deeds, they are ghosts that can take on flesh only when occasions arise that justify our telling them to others.

To strengthen our sense of self the past needs to be rescued and made accessible. Various devices exist to shore up the crumbling landscapes of the past. For example, we can visit the tavern: it provides an opportunity to talk and turn our small adventures into epics, and in some such fashion ordinary lives achieve recognition and even brief glory in the credulous minds of fellow inebriates. Friends depart, but their letters are tangible evidence of their continuing esteem. Relatives die and yet remain present and smiling in the family album. Our own past, then, consists of bits and pieces. It finds a home in the high school diploma, the wedding picture, and the stamped visas of a dogeared passport; in the stringless tennis racket and the much-traveled trunk; in the personal library and the old family home. What objects best image our being? The grandfather clock and the heirloom silverware? The contents of a desk drawer? Books? "A book in one's own library," says the pseudonymous Aristides, "is in a sense a brick in the building of one's being, carrying with it memories, a small block of one's personal intellectual history, associations unsortable in their profusion."[9]

Objects anchor time. They need not, of course, be personal possessions. We can try to reconstruct our past with brief visits to our old neighborhood and the birthplaces of our parents. We can also recapture our personal history by maintaining contact with people who have known us when we were young. Personal possessions are perhaps more important for old people. They are too weary to define their sense of self by projects and action; their social world shrinks and with it the opportunities to proclaim fair deeds; and they may be too fragile to visit places that hold for them fond memories. Personal possessions—old letters and the family settee—remain as accessible comforts, the flavor of times past hovering about them.

Young people live in the future; what they do rather than what they possess defines their sense of selfhood. Yet the young occasionally look back; they can feel nostalgic toward their own short past and proprietary about things. In modern society the teenager, as both his body and his mind undergo rapid change, may have an infirm grasp of who he is. The world seems at times beyond his control. Security lies in routine, in what the teenager perceives to be his own sheltered childhood and in the objects identified with an earlier, more stable phase of life.[10] In general, we may say that whenever a person (young or old) feels that the world is changing too rapidly, his characteristic response is to evoke an idealized and stable past. On the other hand, when a person feels that he himself is directing the change and in control of affairs of importance to him, then nostalgia has no place in his life: action rather than mementos of the past will support his sense of identity.

Some people try hard to recapture the past. Others, on the contrary, try to efface it, thinking it a burden like material possessions. Attachment to things and veneration for the past often go together. A person who likes leather-bound books and oak beams in the ceiling is *ipso facto* an acolyte of history. In contrast, one who disdains possessions and the past is probably a rationalist or a mystic. Rationalism is unsympathetic to clutter. It encourages the belief that the good life is simple enough for the mind to design independently of tradition and custom, and that indeed tradition and custom can cloud the prism of rational thought. Mysticism likewise disdains clutter, material and mental. It declares historical time to be an illusion. Man's essential being belongs to eternity. A mystic frees himself from the burden of material things. He lives in a hermit's cell or by Walden Pond. He is disencumbered of his past.

Societies, like human individuals, differ in their attitudes toward time and place. Nonliterate cultures are, in Lévi-Strauss's word, "cold." Cold societies seek to annul the possible effects of historical events on their equilibrium and continuity. They deny change and try, "with a dexterity we underestimate," to make the status of their development as permanent as possible.[11] The Pygmies of the Congo rain forest have a very shallow sense of time. They lack a creation story; genealogy and even animal life cycles are of little interest. They appear to live wholly in the present. What is there in their environment to remind them of a lengthening past? The rain forest is unchanging. Whatever is made by the Pygmies is made quickly and almost as quickly disintegrates, so that there are few objects that can be handed down from generation to generation as tokens of times gone by.

The Australian aborigines, in comparison, have a much stronger sense of history. Events leading to their present world are recorded in features of the landscape, and each time people pass a particular cleft, cave, or pinnacle they are enabled to recall the deed of an ancestor and culture hero. Still, without a written record and a sophisticated counting system the sense of time cannot be deep. Of the Nuer people in Africa, Evans-Pritchard wrote: "Valid history ends a century ago, and tradition, generously measured, takes us back only ten to twelve generations in lineage structure, and if we are right in supposing that lineage structure never grows, it follows that the distance between the beginning of the world and the present day remains unalterable. . . . How shallow is Nuer time may be judged from the fact that the tree under which mankind came into being was still standing in Western Nuerland a few years ago!"[12]

Among nonliterate peoples, not only the means but the desire to think historically is lacking. The ideal is not development but equilibrium, a state of unvarying harmony. The world as it exists is to be maintained, or restored to pristine perfection. Maturity rather than primitive beginnings is valued. A boy is reborn at the initiation ceremony, which enables him to discard his immature years as he prepares to assume the dignity of manhood. Among such people the fumbling steps toward achievement, including the achieved social order, are readily forgotten. Institutions are sanctioned by timeless myths and an unvarying cosmos. Objects as well as places are venerated because they have power or are associated with beings of power, not because they are old. Antiquarianism is alien to primitive thought.

In the literate Oriental societies of China and Japan the historical sense is well developed. The Chinese are famed for ancestor worship, for keeping dynastic annals, and for deferring to the wisdom of the past. However, the Oriental sense of history differs markedly from that of the Western world in the modern period, that is to say, from the eighteenth century onward. In traditional China the image of an ideal world, in which society conforms to the nature of things, tends to override any sense of history as cumulative change. The constant references to a Golden Age in the past are exhortations to restore harmony to the present in accordance with an idealized model. They call for the return to a former social order and to the rites that sustain it. Their tone is not sentimental or nostalgic. The Chinese do not postulate that the material furnishings of life were more gracious in the past and hence merit the compliment of imitation. What ought to be imitated and perpetuated are the abstract and rather austere rules of social harmony.

The form is more important than the particular substance, which is corruptible. Form can be resurrected whereas the matter of which it consists inevitably decays. In Japan this idea of regeneration explains an ancient Shinto custom. At stated intervals Shinto temples are entirely rebuilt and their furnishings and decorations renewed. The great shrines of Ise in particular, the very center of the religion, are rebuilt every twenty years.[13] In contrast, the great Christian shrines of St. Peter's, Chartres, and Canterbury endure for centuries. The forms change in the long process of construction but the substance, once it is in place, remains unaltered.

Stone is the West's material for building monuments. In China and Japan wood is often used, and wood does not last as long. The Chinese civilization is old but the Chinese landscape offers few man-made structures of great antiquity. Very little that can be seen dates back more than a few centuries. The Great Wall itself, or what can be seen of it, is largely the work of Ming dynasty (A.D. 1366–1644). One of the oldest structures extant in China is the flat-arched An-chi Bridge of Hopei province, which was built between A.D. 605 and 616.[14] The walled city, the hump-backed stone bridge, the rock-and-water garden, the pagoda, and the pavilion have an aura of age and permanence. Like works of nature, they seem changeless. The landscape evinces no clear story line; relics that point to stages of the past are not evident.

History has depth, and time bestows value. These ideas are perhaps more likely to develop in people who live surrounded by artifacts that they know to have taken a long time to make. A great cathedral in the Middle Ages is the result of construction effort sustained over a century and longer. Several human generations can be measured against the steady rise of a monumental edifice. The edifice is a public timepiece. The city in which it is located also has temporal depth objectified in the city's successive walls that accrue like the annual rings of an aged tree (Fig. 22). In China, on the other hand, neither large buildings nor even cities take many years to construct. The Chinese build speedily, and not with an eye to eternity, unless it be that of form. For example, work on Ch'ang-an, the Han capital, began in the spring of 192 B.C. and was completed in the autumn of 190 B.C.[15] Emperor Wen, when he assumed power in A.D. 581, aspired to build a capital on an unprecedented scale. He took up residence in his new city only two years later. The Sui emperors also built an eastern capital, Lo-yang, in less than a year (A.D. 605–606), with a labor force of some two million people.[16] Kublai Khan's Cambaluc was raised from new foundations. A wall girdled the city in 1267. Work began on the main halls and palace in 1273 and was completed early the following year. When Marco Polo arrived in 1275 Cambaluc was only a few years old, yet it already bustled with activity.[17]

The European landscape, unlike the Chinese, is historical, a museum of architectural relics. Prehistoric megaliths, Greek temples, Roman aqueducts, medieval churches, and Renaissance palaces stand in sufficient number to affect the atmosphere of the present scene. Striking changes in architectural style encourage the discerning eye to see history as a long

GROWTH RINGS (SUCCESSIVE WALLS) OF PARIS

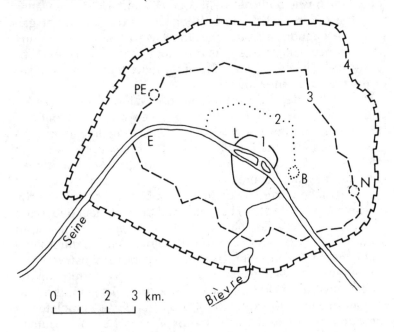

Figure 22. Growth rings (successive walls) of Paris. 1. Wall of Philip Augustus, thirteenth century. 2. Wall at the time of Louis XIV, seventeenth century. 3. Wall of 1840. B. Place de la Bastille. E. Eiffel Tower. L. Louvre. N. Place de la Nation. PE. Place de l'Etoile. Time is made visible in the concentric growth rings of a city. Lucien Gallois, "Origin and growth of Paris," *Geographical Review*, vol. 13, 1923, page 360, figure 12. Reprinted with permission from the American Geographical Society.

chronicle with plots that do not repeat themselves. However, a landscape littered with old buildings does not compel anyone to give it a historical interpretation; one needs a "discerning eye" for such a viewpoint. Until the eighteenth century, time to Europeans had in fact little depth. Remember how in the 1650s Archbishop James Ussher established the creation of the earth itself at 4004 B.C. Remember also that people in the Middle Ages and Renaissance tended to view history primarily as a succession of noble and ignoble deeds and of natural and supernatural events. They showed little awareness of the habits and manners of their forebears living in different periods of the past. They were indeed little aware of the periods themselves. History as the parade of people in fancy costumes and as changing fashions in furniture, so well understood by the modern man who may otherwise be ignorant of history, was alien to medieval thought.[18]

The concept "antique" is modern, as is the idea that old furniture and buildings have a special value bestowed by time and that they should be preserved. Consider the fate of the Colosseum in Rome. Its vast four-storied oval served as a housing site in the Middle Ages. People did not gawk at it; they found shelter in its niches as they would in the caves and on the shelves of a natural scarp. From the fifteenth century on, the Colosseum was robbed of its travertine blocks, which were used in such major buildings as the Palazzo Venezia and St. Peter's. Pope Sixtus V, the great planner of Baroque Rome, had little respect for the artifacts of antiquity; he mined many of the ancient ruins for building material. When toward the end of his short reign he appraised the Colosseum, he did so with the eye of an industrialist rather than that of a historian: he thought the massive structure could be transformed into a colony of workshops for wool-spinners.[19]

Interest in the past waxed with the desire to collect and possess material objects and with the growing prestige of disciplined curiosity. The museum appeared in response to these desires. It began as the private collection of wealthy people who expanded their hoard of familiar art treasures to include oddities, natural and man-made, from widely different parts of the world. At first the collection catered to the pleasure, pride, and enlightenment of a select group only. By the eighteenth century the public was allowed access. At first the collector was not primarily interested in the past; his concern lay in valuable and odd objects, objects that were often considered valuable because odd—odd rather than old. Obviously collection gained interest as the items were labeled and classified. And to the Western mind the simplest taxonomy called for the coordinates of time and place: a coin or a piece of bone belonged to a certain period in the past and came from a certain locality.

In the Age of Enlightenment cultivated Europeans showed increasing fascination with the past, with the idea of development and of memory. As they catalogued items in their museum collections they were led to ponder on the length of the human time span. The new sciences of natural history and geology reminded them that nature's myriad forms had antecedents. In philosophy a major interest of the age was the phenomenon of memory. By remembering, philosophers pointed out, man could escape the purely momentary sensations, the nothingness that lay in wait for him between moments of existence.[20] And what better aid to memory than the tangible evidences of the past—old furniture, old buildings, and museum collections?

The cult of the past, as manifested in the establishment of museums and in the preservation of old buildings, was a type of consciousness that emerged at a certain stage in Europe's history. It has little in common with the fact of being rooted in place. The state of rootedness is essentially subconscious: it means that a people have come to identify themselves with a particular locality, to feel that it is their home and the home of their ancestors. The museum reflects a habit of mind opposed to one that perceives place to be rooted, sacred, and inviolable. The museum, after all, consists wholly of displaced objects. Treasures and oddities are torn from their cultural matrices in different parts of the world and put on pedestals in an alien environment. When London Bridge was dismantled and transported across an ocean and a continent to be rebuilt in the desert of Arizona, the media described the event as a typical example of American folly. It was unique, however, only in scale, for the undertaking reveals an attitude to time and place that is essentially the same as that of Lord Elgin (1766–1841), who removed Athenian marbles to exhibit them in the halls of the British Museum.

The cult of the past calls for illusion rather than authenticity. Ruins in the landscape garden, fashionable for a time in the eighteenth century, made no pretense to being genuine. What mattered was that they provided a mood of time-soaked melancholy. In a museum the complete original artifact is the desideratum, but entire pots are put together from a few fragments and whole animals are re-created from small pieces of bone. The principle for restoring a historic room is similar. Try to obtain the original furnishings. If they cannot be found, antiques resembling the originals may be sought. When antiques are not available, modern reproductions are substituted. An important service of museums is to generate didactic illusions.

Americans of the Revolutionary and post-Independence period wished to deny the European legacy, including the value placed on the past, but they could not be more than partially successful. As a nation born in the eighteenth century America inherited some of Europe's veneration for classical Rome and Greece, as well as Europe's fascination with time and memory. Thomas Jefferson, an iconoclast in some moods, nevertheless designed his university in the classical style, and when he viewed the Blue Ridge landscape his mind was drawn to reflect on its great antiquity.[21] Historical societies soon appeared in the young nation, first in Boston in 1791 and then in New York in 1804. Others followed. In every case their purpose was to collect and preserve documents that would tell the story of their area. Old furniture, tools, and other bric-a-brac were three-dimensional documents that became the core of future museum collections.[22]

When a people deliberately change their environment and feel they are in control of their destiny, they have little cause for nostalgia. Historical societies need not be backward-looking; they may be founded to preserve materials that mark the stages of confident growth and point to the future. When, on the other hand, a people perceive that changes are occurring too rapidly, spinning out of control, nostalgia for an idyllic past waxes strong. In the United States, soon after the centennial celebrations the nostalgic past began to overshadow the past perceived as stages of dynamic growth.[23] Historical societies and museums proliferated to serve both perceptions of time. By the 1960s some 2,500 history museums were open in the United States as against the 274 museums known to be operating on the Indian subcontinent.[24]

Preserving historic buildings, and even whole neighborhoods, is a concern of architect-planners and citizens in both Europe and America. Why preserve? What is the principle behind saving one building rather than another? To simplify the problem these questions raise, look first at the life of a human individual rather than that of a city. A man, let us say, has lived in the same house for many years. By the time he is fifty his house is cluttered with the accumulations of a busy life. They are comfortable mementos of his past, but eventually some of them have to be discarded; they threaten to stand in the way of his present and future projects. He decides to throw much away and keep what is of value to him. He is called upon to evaluate his own past. What does he wish to remember? Evidences of failure, such as rejections slips from publishers and the old copying machine he never learned to use, are quickly junked. A man is not an archivist of his own life, obliged to preserve documents impartially for a future historian to interpret: he wants a commodious house filled with objects that support his sense of self. Valuables are kept, as are old letters and knickknacks that have sentimental worth and do not take up much space. What about the four-poster bed in the guest room? It has been in the family for a long time, it shows good workmanship, but it is also hard on the human spine and leaves little room for closets. Should his guest be made to suffer for his piety?

City authorities and citizens are faced with an essentially similar problem. What facets of the city's past should be preserved? Not the evidences of societal failure, such as old prisons, mental hospitals, and workhouses. These are removed with no regret or second thought on the inviolate nature of history. Art treasures and books are kept. They end up in galleries and libraries. Documents and records are filed away. Such things individually and collectively use little space on the scale of a city. But what about old houses that once belonged to important personages, and malfunctioning department stores that have architectural merit? Unlike precious pictures and books, old buildings occupy much city space and come into conflict with current needs and aspirations.

The passion for preservation arises out of the need for tangible objects that can support a sense of identity. This theme has already been explored. If we turn to the preservationist's reasons for wanting to maintain aspects of the past, they appear to be of three kinds: aesthetic, moral, and morale-boosting. An old edifice, it is argued, should be saved for posterity because it has architectural merit and because it is an achievement of one's forebears. The reason is based on aesthetics, tinged with piety. An old house ought to be preserved because it was once the home of a famous statesman or inventor. Here the appeal is to piety and to the end of building a people's morale, their sense of pride. An old run-down neighborhood should be saved from urban renewal because it seems to satisfy the needs of the local residents, or because, despite a decaying physical environment, it promotes certain human virtues and a colorful style of life. The appeal is to qualities inherent in established ways and to the people's moral right to maintain their distinctive customs against the forces of change.[25]

Why risk change? The past really existed. All that we are we owe to the past. The present also has merit; it is our experiential reality, the feeling point of existence with its inchoate mixture of joy and sorrow. The future, in contrast, is a vision. Many visions go unrealized and some turn into nightmares. A political revolutionary promises us a new earth and may give us chaos or tyranny. An architectural revolutionary promises us a new city and may give us empty lawns and full parking lots. On the other hand, without vision and the desire for change life turns stale; and it is a fact that all creative effort—including the making of an omelette—is preceded by destruction. What future achievement might justify the removal of any urban tissue that still shows signs of life? Planners and citizens, sensitive to past errors, rightly hesitate to sacrifice the present, with all its problems, for a vision of the future that may not be realized. Yet there are striking examples of successful responses to unsought challenges. Cities have repeatedly succumbed to the violence of nature and of war. Thus when fire hollowed out medieval London, an earthquake demolished much of San Francisco, and Nazi bombers flattened Rotterdam, human vision and will were able to overcome disaster. Out of the ruins new cities of no less distinction and greater functionality emerged.[26]

We have examined briefly certain relationships between time and the experience of place. The main points are these: (1) If time is conceived as flow or movement then place is pause. In this view human time is marked by stages as human movement in space is marked by pauses. Just as time may be represented by an arrow, a circular orbit, or the path of a swinging pendulum, so may movements in space; and each representation has its characteristic set of pauses or places. (2) While it takes time to form an attachment to place, the quality and intensity of experience matters more than simple duration. (3) Being rooted in a place is a different kind of experience from having and cultivating a "sense of place." A truly rooted community may have shrines and monuments, but it is unlikely to have museums and societies for the preservation of the past. The effort to evoke a sense of place and of the past is often deliberate and conscious. To the extent that the effort is conscious it is the mind at work, and the mind—if allowed its imperial sway—will annul the past by making it all present knowledge.[27]

14

Epilogue

Human beings, like other animals, feel at home on earth. We are, most of the time, at ease in our part of the world. Life in its daily round is thoroughly familiar. Toast for breakfast is taken for granted, likewise the need to be in the office on time. Skills once learned are as natural to us as breathing. Above all, we are oriented. This is a fundamental source of confidence. We know where we are and we can find our way to the local drugstore. Striding down the path in complete confidence, we are shocked when we miss a step or when our body expects a step where none exists.

Learning is rarely at the level of explicit and formal instruction. The infant acquires a sense of distance by attending to the sound of a human voice that signals the approach of his mother. A child is walked to school a few times and thereafter he can make the trip on his own, without the help of a map; indeed, he is unable to envisage the route. We are in a strange part of town: unknown space stretches ahead of us. In time we know a few landmarks and the routes connecting them. Eventually what was strange town and unknown space becomes familiar place. Abstract space, lacking significance other than strangeness, becomes concrete place, filled with meaning. Much is learned but not through formal instruction.

Epilogue

Nearly all learning is at the subconscious level. Thus we acquire a taste for a certain dish, learn to like a person, appreciate a painting, and grow fond of a neighborhood or resort. Things that were once out of focus for us come into focus, and we perceive them to be individuals and unique. This power to see people and places in their complex particularity is most highly developed in human beings. It is a sign of our superior intelligence, yet we rarely feel the need to apply the power in any systematic way. We claim to know a friend or our hometown well, although we have not done research on either. Even the acquisition of a skill does not always call for explicit instruction. Eskimo children, for example, become hunters by watching adults at work and by doing. We learn to ride a bicycle without a manual of physics; formal knowledge of the balance of forces may even be a handicap.

Routine activity and standard performance do not require analytical thought. When we wish to do something new or to excel, then we need to pause, envision, think. An athlete must of course work at his skill, but his performance will improve if he *thinks* about his movements and tries to perfect them in thought as well as in the field. Thinking and planning help to develop human spatial ability in the sense of agile bodily movements. But far more impressive is the effect of thinking and planning on spatial ability taken in the sense of "conquering space." With the aid of charts and compass (products of thought), human beings have sailed across the oceans; with even more sophisticated instruments they can take leave of the earth itself and fly to the moon.

Analytical thought has transformed our physical and social environment. Evidences of its power are everywhere. We are so impressed that to us "knowing" is practically identical with "knowing about," and Lord Kelvin has gone so far as to say that we do not really know anything unless we can also measure it. Much of human experience is difficult to articulate, however, and we are far from finding devices that measure satisfactorily the quality of a feeling or aesthetic response. What we cannot say in an acceptable scientific language we tend to deny or forget. A geographer speaks as though his knowledge of space and place were derived exclusively from books, maps, aerial photographs, and structured field surveys. He writes as though people were endowed with mind and vision but no other sense with which to apprehend the world and find meaning in it. He and the architect-planner tend to assume familiarity—the fact that we are oriented in space and at home in place—rather than describe and try to understand what "being-in-the-world" is truly like.

A large body of experiential data is consigned to oblivion because we cannot fit the data to concepts that are taken over uncritically from the physical sciences. Our understanding of human reality suffers as a result. Interestingly, this blindness to the depth of experience afflicts the man in the street no less than it does the social scientist. Blindness to experience is in fact a common human condition. We rarely attend to what we know. We attend to what we know about; we are aware of a certain kind of reality because it is the kind that we can easily show and tell. We know far more than what we can tell, yet we almost come to believe that what we can tell is all we know. At a party someone asks, "How do you like Minneapolis?" The typical response is: "It's a good city, a good place to live in, except perhaps for the winter, which seems to last forever." Thus with tired phrases our personal and subtle experiences are misrepresented time and again. Another form of lazy communication is the colored slide show of the family outing. Its effect on captive guests is soporific. To those who have taken the trip each picture may suggest something intimate, such as the feel of the warm sand between the toes, that does not appear on the slide. But to guests the pictures are only pictures, often visual clichés that threaten to march over them in endless platoons.

As social beings and scientists we offer each other truncated images of people and their world. Experiences are slighted or ignored because the means to articulate them or point them out are lacking. The lack is not due to any inherent deficiency in language. If something is of sufficient importance to us we usually find the means to give it visibility. Snow is snow, undifferentiated phenomenon to urban man, but the Eskimo has a dozen words to express it. Feelings and intimate experiences are inchoate and unmanageable to most people, but writers and artists have found ways of giving them form. Literature, for example, is full of precise descriptions of how people live. The academic disciplines themselves yield abundant experiential data that deserve our closer attention.

A rich body of material already exists for the student of environment and man. (And who, in his own way, is not such a student?) For him—that is, for all of us—a basic problem is how to organize this eclectic material. The present essay is *one* attempt to systematize human experiences of space and place. It can claim success if it has made the reader see the range and complexity of experience, and if in addition it has clarified some of the more systematic relationships between and among the wealth of experiential components. But the essay has a still larger purpose, which is that the kinds of questions it poses (if not the answers) enter the debate of environmental design. The discourse of planners and designers must be enlarged to include questions such as these: What connection is there between space awareness and the idea of future time and of goal? What are the links between body postures and personal relationships on the one hand and spatial values and distance relationships on the other? How do we describe "familiarity," that quality of "at homeness" we feel toward a person or place? What kinds of intimate places can be planned, and what cannot—at least, no more than we can plan for deeply human encounters? Are space and place the environmental equivalents of the human need for adventure and safety, openness and definition? How long does it take to form a lasting attachment to place? Is the sense of place a quality of awareness poised between being rooted in place, which is unconscious, and being alienated, which goes with exacerbated consciousness—and exacerbated because it is only or largely mental? How do we promote the visibility of rooted communities that lack striking visual symbols? What is the loss and gain in such promotion?

These questions do not make the life of the social scientist and planner any easier. They make it, for the time being, more difficult by opening up facts that both professionals and non-professionals have found convenient to forget. If we examine certain visionary plans, study certain social surveys, and eavesdrop on the small talk that is the common fare of life, we are likely to discover that whereas the world is enormously complex, human beings and their experiences are simple. The scientist postulates the simple human being for the limited purpose of analyzing a specific set of relationships, and this procedure is entirely valid. Danger occurs when the scientist then naïvely tries to impose his findings on the real world, for he may forget that the simplicity of human beings is an assumption, not a discovery or a necessary conclusion of research. The simple being, a convenient postulate of science and a deliberate paper figure of propaganda, is only too easy for the man in the street—that is, most of us—to accept. We are in the habit of denying or forgetting the real nature of our experiences in favor of the clichés of public speech. And here is the ultimate ambition of this essay, in common with the thrust of humanistic enterprise: to increase the burden of awareness.

NOTES

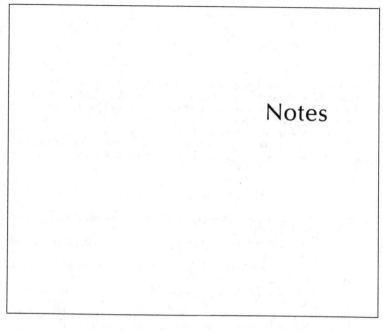

Notes

[1]
Introduction

1. Paul Tillich, *My Search for Absolutes* (New York: Simon and Schuster, 1967), p. 29.
2. Werner Heisenberg, *Physics and Beyond: Encounters and Conversations* (New York: Harper Torchbook, 1972), p. 51.
3. The following recent publications suggest a growing interest in the study of "place" from a variety of humanistic perspectives. John Barrell, *The Idea of Landscape and the Sense of Place, 1730–1840* (Cambridge at the University Press, 1972); Brian Goodey, "The sense of place in British planning: some considerations," *Man-Environment Systems*, vol. 4, no. 4, 1974, pp. 195–202; Linda Graber, *Wilderness as Sacred Space* (Association of American Geographers, Monograph series no. 8, Washington, D.C., 1976); Alan Gussow, *A Sense of Place: The Artist and the American Land* (New York: Seabury, 1974); J. B. Jackson, *Landscapes: Selected Writings*, ed. E. H. Zube (University of Massachusetts Press, 1970); Peirce F. Lewis, "Small town in Pennsylvania," *Annals, Association of American Geographers*, vol. 62, 1972, pp. 323–351; David Ley, *The Black Inner City as Frontier Outpost* (Association of American Geographers, Monograph series no. 7, Washington, D.C., 1974); Lyn H. Lofland, *A World of Strangers: Order and Action in Urban Public Space* (New York: Basic Books, 1973); David Lowenthal, "Past time, present place: landscape and memory," *Geographical Review*, vol. 65, no. 1, 1975, pp. 1–36; Kevin Lynch, *What Time is This Place?* (Cambridge: MIT Press, 1972); D. W. Meinig, "Environmental appreciation: localities as a humane art," *The Western Humanities Review*, vol. 25, 1971, pp. 1–11; C. Norberg-Schulz, *Existence, Space and Architecture* (New York: Praeger, 1971); Kenneth Olwig, "Place, society and the individual in the authorship of St. St. Blicher," in Felix Norgaard, ed.,

Notes

Omkring Blicher 1974 (Denmark: Gyldendal, 1974), pp. 69–114; Edward Relph, *Place and Placelessness* (London: Pion, 1976); Edward H. Spicer, "Persistent cultural systems: a comparative study of identity systems that can adapt to contrasting environments," *Science*, vol. 174, 19 November 1971, pp. 795–800; Mayer Spivak, "Archetypal place," *Architectural Forum*, October 1973, pp. 44–49; Victor Turner, "The center out there: pilgrim's goal," *History of Religions*, vol. 12, no. 3, 1973, pp. 191–230.

[2]
Experiential Perspective

1. Michael Oakeshott, *Experience and Its Modes* (Cambridge at the University Press, 1933), p. 10.
2. Paul Ricoeur, *Fallible Man: Philosophy of the Will* (Chicago: Henry Regnery Co., 1967), p. 127.
3. The German word *erfahren* includes the different meanings of "to find out," "to learn," and "to experience."
4. Susanne K. Langer, *Philosophy in a New Key* (New York: Mentor Book, 1958), p. 85.
5. José Ortega y Gasset, *Man and People* (New York: Norton Library, 1963), pp. 158–159; Julián Marías, *Metaphysical Anthropology: The Empirical Structure of Human Life* (University Park: Pennsylvania State University Press, 1971), p. 40.
6. R. W. Moncrieff, *Odour Preferences* (London: Leonard Hill, 1966), p. 65.
7. *Ibid.*, p. 246.
8. Susanne K. Langer, *Mind: An Essay on Human Feeling* (Baltimore: Johns Hopkins University Press, 1972), vol. 2, pp. 192–193.
9. *Ibid.*, pp. 257–259.
10. Géza Révész, "The problem of space with particular emphasis on specific sensory spaces," *American Journal of Psychology*, vol. 50, 1937, pp. 429–444.
11. Bernard G. Campbell, *Human Evolution: An Introduction to Man's Adaptations* (Chicago: Aldine, 1966), pp. 78, 161–162.
12. William James, *The Principles of Psychology* (New York: Henry Holt, 1918), vol. 2, p. 134.
13. D. M. Armstrong, *Bodily Sensations* (London: Routledge & Kegan Paul, 1962), p. 21.
14. Albert Camus, *Carnet, 1942–1951* (London: Hamish Hamilton, 1966), p. 26.
15. Susanne K. Langer, *Feeling and Form: A Theory of Art* (New York: Charles Scribner, 1953), p. 117.
16. Roberto Gerhard, "The nature of music," *The Score*, no. 16, 1956, p. 7; quoted in Sir Russell Brain, *The Nature of Experience* (London: Oxford University Press, 1959), p. 57.
17. P. H. Knapp, "Emotional aspects of hearing loss," *Psychosomatic Medicine*, vol. 10, 1948, pp. 203–222.
18. James, *Principles of Psychology*, pp. 203–204.
19. *Ibid.*, p. 204.
20. "Those of you who have ever crossed the bay from the Oakland mole to the Ferry Building in San Francisco may include, as I do, a tactual memory of the trip—the touch of the spray and the wind on your face—that com-

bines with the visual image of the bridge and the skyline." George S. Welsh, "The perception of our urban environment," in *Perception and Environment: Foundations of Urban Design*, Institute of Government, University of North Carolina, 1966, p. 6.

[3]
Space, Place, and the Child

1. Ernest G. Schachtel, *Metamorphosis: On the Development of Affect, Perception, Attention, and Memory* (New York: Basic Books, 1959), pp. 287–288, p. 298.
2. J. S. Bruner et al., *Studies in Cognitive Growth* (New York: John Wiley, 1966), p. 2; Wilder Penfield, *The Mystery of the Mind* (Princeton: Princeton University Press, 1975), p. 19.
3. Bing-chung Ling, "A genetic study of sustained visual fixation and associated behavior in the human infant from birth to six months," *Journal of Genetic Psychology*, vol. 61, 1942, pp. 271–272; M. Scaife and J. S. Bruner, "The capacity for joint visual attention in the infant," *Nature*, 24 January 1975, pp. 265–266.
4. B. E. McKenzie and R. H. Day, "Object distance as a determinant of visual fixation in early infancy," *Science*, vol. 178, 1972, pp. 1108–1110.
5. René A. Spitz, *The First Year of Life* (New York: International Universities Press, 1965), p. 64.
6. Jean Piaget, *The Construction of Reality in the Child* (New York: Ballantine Books, 1971), pp. 46–47; Gerald Gratch, "Recent studies based on Piaget's view of object concept development," in Leslie B. Cohen and Philip Salapatek, eds., *Infant Perception: From Sensation to Cognition* (New York: Academic Press, 1975), vol. II, pp. 51–99.
7. Jean Piaget and Bärbel Inhelder, *The Child's Concept of Space* (New York: Norton Library, 1967), p. 5.
8. T. G. R. Bower, "The visual world of infants," *Scientific American*, vol. 215, no. 6, 1966, p. 90; Albert Yonas and Herbert J. Pick, Jr., "An approach to the study of infant space perception," in Cohen and Salapatek, *Infant Perception*, pp. 3–28; Daphne M. Maurer and Charles E. Maurer, "Newborn babies see better than you think," *Psychology Today*, vol. 10, no. 5, 1976, pp. 85–88.
9. Piaget, *The Construction of Reality*, p. 285.
10. Spitz, *The First Year of Life*, p. 176.
11. G. A. Morgan and H. N. Ricciuti, "Infants' response to strangers during the first year," in B. M. Foss, ed., *Determinants of Infant Behavior* (London: Methuen, 1967), p. 263.
12. Eleanor Gibson, *Principles of Perceptual Learning and Development* (New York: Appleton-Century-Crofts, 1969), pp. 319–321.
13. M. J. Konner, "Aspects of the developmental ethology of a foraging people," in N. Blurton Jones, ed., *Ethological Studies of Child Behavior* (Cambridge at the University Press, 1972), p. 297.
14. J. W. Anderson, "Attachment behavior out of doors," in N. Blurton Jones, *Ethological Studies of Child Behavior*, p. 205.
15. *Ibid.*, p. 208.
16. Roman Jakobson, *Child Language Aphasia and Phonological Universals*

(The Hague: Mouton, 1968); quoted in Howard Gardner, *The Quest for Mind* (New York: Vintage Books, 1974), pp. 198–199.

17. Arnold Gesell, F. L. Ilg, and G. E. Bullie, *Vision: Its Development in Infant and Child* (New York: Paul B. Hoeber, 1950), pp. 102, 113, 116.
18. L. B. Ames and J. Learned, "The development of verbalized space in the young child," *Journal of Genetic Psychology*, vol. 72, 1948, pp. 63–84.
19. Piaget and Inhelder, *The Child's Concept of Space*, p. 68, pp. 155–160, p. 20.
20. D. R. Olson, *Cognitive Development: The Child's Acquisition of Diagonality* (New York: Academic Press, 1970).
21. Jean Piaget, *The Child and Reality* (New York: Viking Compass Edition, 1974), p. 19, 86. See also Roger A. Hart and Gary T. Moore, "The development of spatial cognition: a review," in Roger M. Downs and David Stea, eds., *Image and Environment* (Chicago: Aldine, 1973), pp. 246–288.
22. Piaget and Inhelder, *The Child's Concept of Space*, pp. 379, 389.
23. *Ibid.*, p. 49.
24. J. M. Blaut and David Stea, "Studies of geographic learning," *Annals, Association of American Geographers*, vol. 61, no. 2, 1971, pp. 387–393, and David Stea and J. M. Blaut, "Some preliminary observations on spatial learning in school children," in Downs and Stea, *Image and Environment*, pp. 226–234.
25. Susan Isaacs, *Intellectual Growth in Young Children* (New York: Harcourt and Brace, 1930), p. 37.
26. Ruth M. Beard, *An Outline of Piaget's Developmental Psychology* (New York: Mentor Book, 1972), pp. 109–110.
27. Gesell et al., *Vision*, p. 126.
28. John Holt writes: "The courage of little children (and not them alone) rises and falls, like the tide—only the cycles are in minutes, or even seconds. We can see this vividly when we watch infants of two or so, walking with their mothers, or playing in a playground or park. Not long ago I saw this scene in the Public Garden in Boston. The mothers were chatting on a bench while the children roamed around. For a while they would explore boldly and freely, ignoring their mothers. Then, after a while, they would use up their store of courage and confidence, and run back to their mothers' sides, and cling there for a while, as if to recharge their batteries. After a moment or two of this they were ready for more exploring, and so they went, out, then came back, and then ventured out again." In *How Children Learn* (New York: Dell Publishing Co., 1970), p. 101.
29. Gesell et al., *Vision*, p. 121.
30. Ames and Learned, "The development of verbalized space," pp. 72, 75.
31. F. J. Estvan and E. W. Estvan, *The Child's World: His Social Perception* (New York: G. P. Putnam's, 1959), pp. 21–76.
32. Jean Piaget, *The Child's Conception of the World* (Totowa, New Jersey: Littlefield, Adams, 1969), pp. 352–354.
33. Susan Isaacs, "Property and possessiveness," in Toby Talbot, ed., *The World of the Child* (Garden City, New York: Anchor Books, 1968), pp. 255–265.
34. Robert Coles, *Migrants, Sharecroppers, Mountaineers* (Boston: Atlantic-Little, Brown, 1972), p. 67.
35. S. Honkavaara, *The Psychology of Expression*, British Journal of Psychology Monograph Supplements, no. 32, 1961, pp. 41–42, p. 45; Howard

Gardner and Ellen Winner, "How children learn: three stages of understanding art," *Psychology Today*, vol. 9, no. 10, 1976, pp. 42–45, p. 74.

[4]
Body, Personal Relations, and Spatial Values

1. Immanuel Kant, "On the first ground of the distinction of regions in space," in *Kant's Inaugural Dissertation and Early Writings on Space*, trans. John Handyside (Chicago: Open Court, 1929), pp. 22–23. See also J. A. May, *Kant's Concept of Geography and Its Relation to Recent Geographical Thought*, University of Toronto Department of Geography Research Publication no. 4 (University of Toronto Press, 1970), pp. 70–72.
2. Arnold Gesell and Catharine S. Amatruda, *Developmental Diagnosis* (New York: Harper & Row, 1947), p. 42.
3. E. W. Straus, *Phenomenological Psychology* (New York: Basic Books, 1966), p. 143.
4. E. R. Bevan, *Symbolism and Belief* (London: George Allen and Unwin, 1938), p. 48.
5. Michael Young and Peter Willmott, *The Symmetrical Family* (New York: Pantheon Books, 1973), pp. 44–45.
6. René Guénon, "L'Idée du centre dans la tradition antique," in *Symboles fondamentaux de la science sacrée* (Paris: Gallimard, 1962), pp. 83–93; Paul Wheatley, "The symbolism of the center," in *The Pivot of the Four Quarters* (Chicago: Aldine, 1971), pp. 428–436.
7. Uno Holmberg, "Siberian mythology," in J. A. MacCulloch, ed., *Mythology of All the Races* (Boston: Marshall Jones, 1927), vol. 4, p. 309.
8. Bevan, *Symbolism and Belief*, p. 66.
9. A. J. Wensinck, "Ka'ba" in *The Encyclopaedia of Islam* (Leiden: Brill, 1927), vol. 2, p. 590.
10. John Wesley, *A Survey of the Wisdom of God in the Creation* (London: 1809), vol. 3, p. 11.
11. Marcel Granet, "Right and left in China," in R. Needham, ed., *Right & Left: Essays on Dual Symbolic Classification* (Chicago: University of Chicago Press, 1973), p. 49.
12. Ervin Goffman, *The Presentation of Self in Everyday Life* (Garden City, N.Y.: Doubleday Anchor, 1959), p. 123.
13. A. F. Wright, "Symbolism and function: reflections on Changan and other great cities," *Journal of Asian Studies*, vol. 24, 1965, p. 671.
14. D. C. Munro and G. C. Sellery, *Medieval Civilizations: Selected Studies from European Authors* (New York: The Century Co., 1910), pp. 358–361. With regard to Asian traditions, Paul Wheatley wrote: "The city gates, where power generated at the *axis mundi* flowed out from the confines of the ceremonial complex towards the cardinal points of the compass, possessed a heightened symbolic significance which, in virtually all Asian urban traditions, was expressed in massive constructions whose size far exceeded that necessary for the performance of their mundane functions of granting access and affording defense." "The symbolism of the center," p. 435.
15. Documented in Needham, ed., *Right & Left*.

16. Robert Hertz, *Death and the Right-Hand* (Glencoe, Illinois: Free Press, 1960), pp. 100–101.
17. A. C. Kruyt, "Right and left in central Celebes," in Needham, ed., *Right & Left*, pp. 74–75.
18. J. Chelhod, "Pre-eminence of the right, based upon Arabic evidence," in Needham, ed., *Right & Left*, pp. 246–247.
19. James Littlejohn, "Temne right and left: an essay on the choreography of everyday life," in Needham, ed., *Right & Left*, p. 291.
20. Granet, "Right and left in China," pp. 43–58.
21. Henri Frankfort, H. A. Frankfort, John A. Wilson, and Thorkild Jacobsen, *Before Philosophy* (Baltimore: Penguin, 1951), pp. 45–46.
22. Carl H. Hamburg, *Symbol and Reality* (The Hague: Martinus Nijhoff, 1970), p. 98.
23. D. Westermann, *A Study of the Ewe Language* (London: Oxford University Press, 1930), pp. 52–55.
24. Ernst Cassirer, *The Philosophy of Symbolic Forms* (New Haven: Yale University Press, 1953), pp. 206–207.
25. Maurice Merleau-Ponty, *Phenomenology of Perception* (London: Routledge & Kegan Paul, 1962), p. 101.
26. Jean-Paul Sartre, "The body," in Stuart F. Spicker, ed., *The Philosophy of the Body* (Chicago: Quadrangle Books, 1970), p. 227.
27. *Notes and Queries in Anthropology*, Committee of the Royal Anthropological Institute (London: Routledge & Kegan Paul, 1951), p. 197.
28. To the Temne of Sierra Leone, "The size of a farm . . . is arrived at by estimating the number of bags of rice it ought to produce. . . . When men hire themselves out to hoe for a farmer, the farmer and the labourer agree on an area which the labourer should complete in a day's work. The day's work however consists of completing the area." J. Littlejohn, "Temne space," *Anthropological Quarterly*, vol. 36, 1963, p. 4.
29. R. H. Codrington, *The Melanesian Languages* (Oxford: Clarendon Press, 1885), pp. 164–165; see also pp. 103–105.
30. Franz Boas, "Kwakiutl," in Franz Boas, ed., *Handbook of American Indian Languages* (Smithsonian Institution, Washington, D.C.: Government Printing Office, 1911), Bulletin 40, part 1, p. 445.
31. *Ibid.*, p. 446.
32. Cassirer, *The Philosophy of Symbolic Forms*, p. 213.
33. John R. Swanton, "Tlingit," in Boas, ed., *Handbook of American Indian Languages*, p. 172.
34. Waldemar Bogoras, "Chukchee," in F. Boas, ed., *Handbook of American Indian Languages* (Smithsonian Institution, Washington, D.C.: Government Printing Office, 1922), Bulletin 40, part 2, p. 723.
35. "I asked [Bertrand Russell—95 years old] how one of his grandchildren was getting on. He didn't at first hear; and Edith said 'Oh she's been doing this and that.' Bertie caught this and said ruefully, 'Mostly *that*!' We speculated as to why, in such verbal pairs, the second is always worse than the first." Rupert Craw‐hay-Williams, *Russell Remembered* (London: Oxford University Press, 1970), p. 152.
36. Stephen A. Erickson, "Language and meaning," in James M. Edie, ed., *New Essays in Phenomenology* (Chicago: Quadrangle Books, 1969), pp. 45–46.

[5]
Spaciousness and Crowding

1. Recent literature on social space and the human psychology of crowding has moved away from naïve inferences based on observations of animal behavior under laboratory conditions. See Irwin Altman, *The Environment and Social Behavior* (Monterey, California: Brooks, Cole Co., 1975); the special issue on "Crowding in real environments," Susan Saegert, ed., in *Environment and Behavior*, vol. 7, no. 2, 1975; Aristide H. Esser, "Experiences of crowding," *Representative Research in Social Psychology*, vol. 4, 1973, pp. 207–218; Charles S. Fischer, Mark Baldassare, and Richard J. Ofshe, "Crowding studies and urban life: a critical review," *Journal of American Institute of Planners*, vol. 43, no. 6, 1975, pp. 406–418; Gunter Gad, "'Crowding' and 'pathologies': some critical remarks," *The Canadian Geographer*, vol. 17, no. 4, 1973, pp. 373–390.
2. Studs Terkel, *Working* (New York: Pantheon, 1974), pp. 385–386.
3. Antoine de Saint-Exupéry, *Wind, Sand, and Stars* (Harmondsworth: Penguin Books, 1966), p. 24.
4. For an extensive analysis of landscape and landscape painting into the categories of "prospect" (space) and "refuge" (place) see Jay Appleton, *The Experience of Landscape* (London: John Wiley, 1975); Edoardo Weiss, *Agoraphobia in the Light of Ego Psychology* (New York: Grune & Stratton, 1964), p. 52, 65. Psychiatrists no longer distinguish sharply between the person who fears open spaces and the person who fears tight small spaces. "The agoraphobic is also likely to be claustrophobic, be afraid of fainting, dying or going mad or losing control." Isaac M. Marks, *Fears and Phobias* (New York: Academic Press, 1969), p. 120.
5. Raymond Firth, *We, the Tikopia* (London: George Allen & Unwin, 1957), p. 19.
6. Burton Watson, *Chinese Lyricism: Shih Poetry from the Second to the Twelfth Century* (New York: Columbia University Press, 1971), p. 21.
7. Maxim Gorky, "On the Russian peasantry," quoted in Jules Koslow, *The Despised and the Damned: The Russian Peasant through the Ages* (New York: Macmillan, 1972), p. 35.
8. Martin Heidegger, "Art and space," *Man and World*, vol. 6, no. 1, 1973, pp. 3–8.
9. Michael Sullivan, *The Birth of Landscape Painting in China* (Berkeley: University of California Press, 1962); Edward H. Schafer, *The Vermilion Bird: The Images of the South* (Berkeley: University of California Press, 1967), pp. 120–122.
10. John F. A. Sawyer, "Spaciousness," *Annual of the Swedish Theological Institute*, vol. 6, 1967–68, pp. 20–34.
11. Ervin Goffman, *Behavior in Public Places* (New York: The Free Press, 1966), p. 15.
12. Aristide H. Esser, *Behavior and Environment: The Use of Space by Animals and Men* (New York: Plenum Press, 1971), p. 8.
13. Mary McCarthy, *The Writing on the Wall* (New York: Harcourt, Brace & World, 1970), p. 203.
14. Jules Henry, *Jungle People: A Kaingáng Tribe of the Highlands of Brazil* (New York: J. J. Augustin, 1941), pp. 18–19.
15. Patricia Draper, "Crowding among hunter-gatherers: the !Kung

Bushmen," *Science*, vol. 182, 19 October 1973, pp. 301–303. For another example of natural crowding without adverse effect, see Albert Damon, "Human ecology in the Solomon Islands: biomedical observations among four tribal societies," *Human Ecology*, vol. 2, no. 3, 1974, pp. 191–215.

16. Alvin L. Schorr, "Housing and its effects," in Harold M. Proshansky, William H. Ittelson, and Leanne G. Rivlin, *Environmental Psychology* (New York: Holt, Rinehart and Winston, 1970), p. 326.

17. The art historian Bernard Berenson wrote: "An Italian crowd is delightful. It does not swear, and it does not use its elbows. To be in the midst of it is truly to be taking *un bain de multitude. . . .*" In *The Bernard Berenson Treasury*, selected and edited by Hanna Kiel (New York: Simon and Schuster, 1962), p. 58.

18. *The New York Times*, Sunday, July 29, 1973, p. 38.

19. In *Doctor Zhivago*; quoted by Edmund Wilson, "Legend and symbol in Doctor Zhivago," in *The Bit between My Teeth* (London: W. H. Allen, 1965), p. 464.

20. O. F. Bollnow, "Lived-space," in Nathaniel Lawrence and Daniel O'Connor, *Readings in Existential Phenomenology* (Englewood Cliffs: Prentice-Hall, 1967), pp. 178–186.

21. Richard Hoggart, *The Uses of Literacy* (New York: Oxford University paperback, 1970), p. 34.

22. Irwin Altman, "Privacy: a conceptual analysis," *Environment and Behavior*, vol. 8, no. 1, 1976, pp. 7–29.

23. Robert Roberts, like Hoggart, comes out of a working-class background. His picture of working-class life is appreciably more somber than that of Hoggart. Robert Roberts, *The Classic Slum: Salford Life in the First Quarter of the Century* (Manchester: Manchester University Press, 1971).

24. Haim Schwarzbaum, "The overcrowded earth," *Numen*, vol. 4, January 1957, pp. 59–74.

25. Knud Rasmussen, *The Intellectual History of the Iglulik Eskimos*, Report of the 5th Thule Expedition, The Danish Expedition to Arctic North America, vol. 7, 1929, pp. 92–93.

[6]
Spatial Ability, Knowledge, and Place

1. "We cannot learn to keep our balance on a bicycle by trying to follow the explicit rule that, to compensate for an imbalance, we must force our bicycle into a curve—away from the direction of the imbalance—whose radius is proportional to the square of the bicycle's velocity over the angle of imbalance. Such knowledge is totally ineffectual unless it is known tacitly, that is, unless it is known subsidiarily—unless it is simply dwelt in." Michael Polanyi and Harry Prosch, *Meaning* (Chicago: University of Chicago Press, 1975), p. 41.

2. I have explored this theme in "Images and mental maps," *Annals, Association of American Geographers*, vol. 65, no. 2, 1975, pp. 205–213.

3. G. G. Luce and Julius Segal, *Sleep* (New York: Coward-McCann, 1966), p. 134.

4. Nathaniel Kleitman, *Sleep and Wakefulness* (Chicago: University of Chicago Press, 1963), p. 282.

5. Griffith Williams, "Highway hypnosis: an hypothesis," *International Journal of Clinical and Experimental Hypnosis*, vol. 11, no. 3, 1963, p. 147.
6. L. A. Pechstein, "Whole vs. part methods in motor learning," *Psychological Monograph*, vol. 33, no. 99, 1917, p. 30; quoted by K. S. Lashley, "Learning: I. Nervous mechanisms in learning," in Carl Murchison, ed., *The Foundations of Experimental Psychology* (Worcester: Clark University Press, 1929), p. 535.
7. Warner Brown, "Spatial integration in human maze," *University of California Publications in Psychology*, vol. 5, no. 6, 1932, pp. 123–134.
8. *Ibid.*, p. 128.
9. *Ibid.*, p. 124.
10. D. O. Hebb, *The Organization of Behavior: A Neuropsychological Theory* (New York: John Wiley, 1949), p. 136.
11. Robert Edgerton, *The Cloak of Competence* (Berkeley: University of California Press, 1967), p. 95.
12. Alan Richardson, *Mental Imagery* (London: Routledge & Kegan Paul, 1969), p. 56; Richard M. Suinn, "Body thinking: psychology for Olympic champs," *Psychology Today*, vol. 10, no. 2, 1976, pp. 38–43.
13. J. A. Leonard and R. C. Newman, "Spatial orientation in the blind," *Nature*, vol. 215, no. 5108, 1967, p. 1414.
14. J. McReynolds and P. Worchel, "Geographic orientation in the blind," *Journal of General Psychology*, vol. 51, 1954, p. 230, 234.
15. H. R. De Silva, "A case of a boy possessing an automatic directional sense," *Science*, vol. 73, 1931, pp. 393–394. Do the Chinese have an unusually developed sense of direction? "In China, when one wishes to have a table moved to a different part of one's room, one does not tell the servant to shift it to his right or his left, but to 'move it a little east' or west . . . even if it is a matter of only two or three inches." Derk Bodde, "Types of Chinese categorical thinking," *Journal of the American Oriental Society*, vol. 59, 1939, p. 201.
16. John Nance, *The Gentle Tasaday* (New York: Harcourt Brace Jovanovich, 1975), pp. 21–22.
17. H. D. Hutorowicz, "Maps of primitive peoples," *Bulletin, American Geographical Society*, vol. 43, 1911, pp. 669–679; C. E. LeGear, "Map making by primitive peoples," *Special Libraries*, vol. 35, no. 3, 1944, pp. 79–83.
18. Hutorowicz, "Maps of primitive peoples," p. 670.
19. John W. Berry, "Temne and Eskimo perceptual skills," *International Journal of Psychology*, vol. 1, 1966, pp. 207–229.
20. Edmund S. Carpenter, "Space concepts of the Aivilik Eskimo," *Explorations*, vol. 5, 1955, p. 140.
21. *Ibid.*, p. 138.
22. Berry, "Temne and Eskimo perceptual skills." See also Beatrice Whiting's discussion on "Differences in child rearing between foragers and nonforagers" in Richard B. Lee and Irven de Vore, eds., *Man the Hunter* (Chicago: Aldine, 1968), p. 337.
23. David Lewis, *We, the Navigators* (Honolulu: The University Press of Hawaii, 1972), pp. 17–18.
24. Thomas Gladwin, *East Is a Big Bird*. (Cambridge: Harvard University Press, 1970) pp. 17–18.
25. *Ibid.*, p. 56.

216

Notes

26. Lewis, *We, the Navigators*, p. 87.
27. Gladwin, *East Is a Big Bird*, p. 129.
28. *Ibid.*, p. 131.
29. *Ibid.*, p. 34; M. Levison, R. Gerard Ward, and J. W. Webb, *The Settlement of Polynesia: A Computer Simulation* (Minneapolis: University of Minnesota Press, 1973), pp. 62–64.

[7]
Mythical Space and Place

1. For Northwest Passage, see John K. Wright, "The open Polar Sea," in *Human Nature in Geography* (Cambridge: Harvard University Press, 1966), pp. 89–118. "As the exploration of North America continued and the various expeditions failed to discover the much desired 'waterway to Cathay,' the strait theory began to lose prestige in some quarters. However, the idea of a water passage to the East was not dispelled though it was changed in form. Now, instead of a broad water passage to the north of the continent, a river was envisioned that would traverse this great area. This view is quite interestingly shown in the note given to the Jamestown colonists in 1607 advising them to settle on a navigable river, 'that which bendeth most towards the N.W., for that way you shall soonest find the other sea.'" In G. G. Cline, *Exploring the Great Basin* (Norman: University of Oklahoma Press, 1963), p. 21. For terrestrial paradise, see Henri Baudet, *Paradise on Earth* (New Haven: Yale University Press, 1965), and Carolly Erickson, *The Medieval Vision: Essays in History and Perception* (New York: Oxford University Press, 1976), pp. 3–8. I wish to thank Ivor Winton for reading this chapter critically.
2. T. A. Ryan and M. S. Ryan, "Geographical orientation," *American Journal of Psychology*, vol. 53, 1940, pp. 204–215.
3. Thomas Gladwin, *East Is a Big Bird: Navigation and Logic on Puluwat Atoll* (Cambridge: Harvard University Press, 1970), pp. 17, 132.
4. A. Irving Hallowell, *Culture and Experience* (Philadelphia: University of Pennsylvania Press, 1955), p. 187.
5. *Ibid.*, pp. 192–193.
6. Victor W. Turner, "Symbols in African ritual," *Science*, vol. 179, 16 March 1973, p. 1104. In this article Turner notes how the complex cosmologies of West Africa differ from the relatively simple myths of Central Africa, and offers some interpretations.
7. *Ibid.* See Geneviève Calame-Griaule, *Ethnologie et Langage: le parole chez les Dogons* (Paris: Gallimard, 1965), pp. 27–28.
8. For Chinese folk belief concerning nature, see E. Chavannes, *Le T'ai Chan* (Paris: Ernest Leroux, 1910).
9. William Caxton, *Mirrour of the World*, ed. Oliver H. Prior (London: Early English Text Society, 1913), p. 109. First published in 1481.
10. Sir Walter Ralegh, *The History of the World*, book 1, chap. 2, sec. 5, in *The Works of Sir Walter Ralegh* (Oxford: 1829), p. 59.
11. Christopher Packe, *A Dissertation upon the Surface of the Earth* (London: 1737), pp. 4–5.
12. Ernst Cassirer, *The Individual and the Cosmos in Renaissance Philosophy* (Oxford: Clarendon Press, 1963), p. 110.

13. Leonard Barkan, *Nature's Work of Art: The Human Body as Image of the World* (New Haven: Yale University Press, 1975), p. 23.
14. Gilbert Cope, *Symbolism in the Bible and the Church* (London: SCM Press, 1959), pp. 63–64.
15. See, for example, Michael Coe, "A model of ancient community structure in the Maya lowlands," *Southwestern Journal of Anthropology*, vol. 21, 1965, pp. 97–113; Robert Fuson, "The orientation of Mayan ceremonial centers," *Annals, Association of American Geographers*, vol. 59, 1969, pp. 494–511; John Ingham, "Time and space in ancient Mexico: the symbolic dimensions of clanship," *Man*, vol. 6, no. 4, 1971, pp. 615–629; Joyce Marcus, "Territorial organization of the Lowland Classic Maya," *Science*, vol. 180, no. 4089, 1973, pp. 911–916; John G. Neihardt, *Black Elk Speaks* (Nebraska: Bison Book, 1961), p. 2; Alfonso Ortiz, "Ritual drama and the Pueblo world view," in *New Perspectives on the Pueblos* (Albuquerque: University of New Mexico Press, 1972), p. 142; Leslie A. White, "The Pueblo of Santa Ana, New Mexico," *American Anthropological Association*, Memoir 60, vol. 44, no. 4, 1942, pp. 80–84.
16. Henri Frankfort, H. A. Frankfort, John A. Wilson, and Thorkild Jacobsen, *Before Philosophy* (Baltimore: Penguin, 1951); Paul Wheatley, *City as Symbol* (London: H. K. Lewis, 1969), pp. 17–21; Werner Müller, *Die heilige Stadt* (Stuttgart: Kohlhammer, 1961); Alfred Forke, *The World-Conception of the Chinese* (London: Arthur Probsthain, 1925); Marcel Granet, *La Pensée Chinoise* (Paris: Albin Michel, 1934), especially the section "Le microcosme," pp. 361–388; Justus M. van der Kroef, "Dualism and symbolic antithesis in Indonesian society," *American Anthropologist*, vol. 56, 1954, pp. 847–862. For a survey of the world's cosmic systems based on the cardinal points and color, see Karl A. Nowotny, *Beiträge zur Geschichte des Weltbildes: Farben und Weltrichtungen*, Wiener Beiträge zur Kulturgeschichte und Linguistik, vol. 17 (1969), Verlag Ferdinand Berger & Söhne, Horn-Wien, 1970. I wish to thank Stephen Jett for this reference.
17. G. W. Stow, *The Native Races of South Africa* (London: Sonnenschein, 1910), p. 43.
18. White, "The Pueblo of Santa Ana, New Mexico," p. 80.
19. Hallowell, *Culture and Experience*, p. 191.
20. *Ibid.*, p. 199.
21. Nelson I. Wu, *Chinese and Indian Architecture* (New York: Braziller, 1963), p. 12.
22. Nowotny, *Beiträge zur Geschichte des Weltbildes*.
23. Franz Cumont, *Astrology and Religion among the Greeks and Romans* (New York: Dover, 1960), p. 67.
24. Vincent Scully, *The Earth, the Temple, and the Gods* (New Haven: Yale University Press, 1962), p. 44.
25. Ernest Brehaut, *An Encyclopedist of the Dark Ages: Isidore of Seville* (Columbia University Studies in History, Economics, and Public Law), vol. 48, 1912, pp. 61–62.
26. *Ibid.*, pp. 238–239.
27. Cope, *Symbolism in the Bible*, pp. 242–243.
28. See, for example, the Ebsdorf world map (ca. 1235), reproduced in Leo Bagrow, *History of Cartography* (Cambridge: Harvard University Press, 1964), plate E.

Notes

29. Donald J. Munro, *The Concept of Man in Early China* (Stanford: Stanford University Press, 1969), p. 41.
30. D. R. Dicks, "The KΛIMATA in Greek geography," *Classical Quarterly*, vol. 5, 1955, p. 249.
31. Clarence J. Glacken, *Traces on the Rhodian Shore* (Berkeley: University of California Press, 1967).
32. Geoffrey Lloyd, "Right and left in Greek philosophy," in Needham, ed., *Right & Left*, p. 177.
33. Loren Baritz, "The idea of the West," *American Historical Review*, vol. 66, no. 3, 1961, pp. 618–640.
34. Merrill Jensen, ed., *Regionalism in America* (Madison: University of Wisconsin Press, 1965); Wilbur Zelinsky, *The Cultural Geography of the United States* (Englewood Cliffs: Prentice-Hall, 1973).
35. Leslie A. Fiedler, *The Return of the Vanishing American* (New York: Stein & Day, 1968), pp. 16–22.
36. J. B. Jackson, *American Space: The Centennial Years 1865–1876* (New York: Norton, 1970), p. 58.
37. Mircea Eliade, *Images and Symbols* (New York: Sheed and Ward, 1969), p. 39.

[8]
Architectural Space and Awareness

1. Karl von Frisch, *Animal Architecture* (New York: Harcourt Brace Jovanovich, 1974).
2. See Christopher Alexander, *Notes on the Synthesis of Form* (Cambridge: Harvard University Press, 1964). On the stability of certain folk architectural forms, Alexander gives the following references: L. G. Bark, "Beehive dwellings of Apulia," *Antiquity*, vol. 6, 1932, p. 410; Werner Kissling, "House tradition in the Outer Hebrides," *Man*, vol. 44, 1944, p. 137; and H. A. and B. H. Huscher, "The hogan builders of Colorado," *Southwestern Lore*, vol. 9, 1943, pp. 1–92.
3. Pierre Deffontaines, *Géographie et religions* (Paris: librarie Gallimard, 1948); Mircea Eliade, *The Sacred and the Profane* (New York: Harper & Row, 1961); Lord Raglan, *The Temple and the House* (London: Routledge & Paul, 1964). On the animal and human sacrifices in the building of a Royal City, see T. K. Chêng, *Shang China* (Toronto: University of Toronto Press, 1960), p. 21.
4. John Harvey, *The Medieval Architect* (London: Wayland Publishers, 1972), p. 97.
5. *Ibid.*, p. 26.
6. P. du Colombier, *Les Chantiers des cathédrales* (Paris: J. Picard, 1953), p. 18; quoted in Adolf Katzenellenbogen, *The Sculptural Programs of Chartres Cathedral* (New York: Norton, 1964), p. vii.
7. Leopold von Ranke, *Ecclesiastical and Political History of the Popes during the Sixteenth and Seventeenth Centuries* (London: J. Murray, 1840), vol. 1, book 4, sec. 8; quoted by Geoffrey Scott, *The Architecture of Humanism* (New York: Charles Scribner's, 1969), pp. 112–113.
8. For the courtyard house in Mesopotamia see C. L. Woolley, "The excavations at Ur 1926–1927," *The Antiquaries Journal* (London), vol. 7, 1927, pp.

Notes

387–395. For a summary of changes in the style of the house in the ancient Near East see S. Giedion, *The Eternal Present* (New York: Pantheon Books, 1964), pp. 182–189. For the evolution of house shape from oval to rectangular in ancient Egypt see Alexander Badawy, *Architecture in Ancient Egypt and the Near East* (Cambridge: MIT Press, 1966), pp. 10–14. For changes in house type from the round semi-subterranean dwelling to the courtyard pattern in Boeotia see Bertha Carr Rider, *Ancient Greek Houses* (Chicago: Argonaut, 1964), pp. 42–68. On Crete, however, Neolithic houses were predominantly rectangular. See D. S. Robertson, *Greek and Roman Architecture* (Cambridge at the University Press, 1969), p. 7. For the development of the Chinese house since Neolithic times see K. C. Chang, *The Archaeology of Ancient China* (New Haven: Yale University Press, 1968), and Andrew Boyd, *Chinese Architecture and Town Planning, 1500 B.C.–A.D. 1911* (Chicago: University of Chicago Press, 1962). For prehistoric Mexico see Marcus C. Winter, "Residential patterns at Monte Alban, Oaxaca, Mexico," *Science*, vol. 186, no. 4168, 1974, pp. 981–986.

9. Neil Harris, "American space: spaced out at the shopping center," *The New Republic*, vol. 173, no. 24, 1975, pp. 23–26.

10. S. Giedion, *The Eternal Present*, particularly "Supremacy of the vertical," pp. 435–492.

11. Wright Morris, *The Home Place* (New York: Charles Scribner's, 1948), pp. 75–76.

12. S. Giedion, *Architecture and the Phenomena of Transition* (Cambridge: Harvard University Press, 1971), pp. 144–255. On the importance and influence of the Pantheon see William L. MacDonald, *The Pantheon: Design, Meaning, and Progeny* (Cambridge: Harvard University Press, 1976).

13. "Open, outdoor space, without limiting contours of hills or shore lines, is many times larger than the hugest edifice, yet the sense of vastness is more likely to beset one upon entering a building; and there it is clearly an effect of pure forms." Susanne K. Langer, *Mind: An Essay on Human Feeling* (Baltimore: Johns Hopkins University Press, 1967), vol. 1, p. 160.

14. Black Elk, the Oglala Sioux, sees the circle and circular processes everywhere in nature as well as in the human world. John G. Neihardt, *Black Elk Speaks* (Lincoln: University of Nebraska Press, 1961), pp. 198–200.

15. Clark E. Cunningham, "Order in the Atoni House," *Bijdragen Tet De Taaland-En Volkekunde*, vol. 120, 1964, pp. 34–68.

16. P. Suzuki, *The Religious System and Culture of Nias, Indonesia* (The Hague: Ph.D. dissertation, Leiden University, 1959), pp. 56ff.; summarized in Douglas Fraser, *Village Planning in the Primitive World* (New York: George Braziller, n.d.), pp. 36–38.

17. Colin M. Turnbull, *Wayward Servants* (London: Eyre & Spottiswode, 1965), p. 200.

18. The teaching function of the cathedral is a theme developed in Emile Mâle, *The Gothic Image* (New York: Harper Torchbooks, 1958).

19. See Patrick Nuttgens, "The metaphysics of light," in *The Landscape of Ideas* (London: Faber & Faber, 1972), pp. 42–60. Otto von Simson wrote: "This attitude [of the Middle Ages] toward sacred architecture differs widely from our own. . . . The simplest way of defining this difference is to recall the changed meaning and function of the symbol. For us the

symbol is an image that invests physical reality with poetical meaning. For medieval man, the physical world as we understand it has no reality except as a symbol. But even the term 'symbol' is misleading. For us the symbol is the subjective creation of poetic fancy; for medieval man what we would call symbol is the only objectively valid definition of reality. . . . Maximus the Confessor . . . defines what he calls 'symbolic vision' as the ability to apprehend within the objects of sense perception the invisible reality of the intelligible that lays beyond them." Simson, *The Gothic Cathedral* (New York: Pantheon Books, 1962), pp. xix–xx.

20. A. Rapoport, "Images, symbols and popular design," *International Journal of Symbology*, vol. 4, no. 3, 1973, pp. 1–12; Marc Treib, "Messages in the interstices: symbols in the urban landscape," *Journal of Architectural Education*, vol. 30, no. 1, 1976, pp. 18–21.

21. Scott, *The Architecture of Humanism*, p. 50.

[9]
Time in Experiential Space

1. Langer, *Feeling and Form*, p. 112.

2. Stephen Shapiro and Hilary Ryglewicz state the relationship between space and time as follows: "People who value neatness are often fond of schedules; the ordering of both space and time enhances their feeling of security. People who are 'loose' with space are often 'loose' with time. . . . We can feel interruptions of our private time much as we feel invasions of our private space. 'Safe space' for the self often means a place, such as home or study room, where one's time is safe from interruption; while 'safe time' often means the time spent in a special place or with a special person." In *Feeling Safe* (Englewood Cliffs, N.J.: Prentice-Hall, 1976), p. 102.

3. Colin M. Turnbull, *The Forest People* (London: Chatto & Windus, 1961), p. 223, 227; "The legends of the BaMbuti," *Journal of the Royal Anthropological Institute*, vol. 89, 1959, pp. 45–60; "The MButi Pygmies: an ethnographic survey," *Anthropological Papers*, The American Museum of Natural History, vol. 50, part 3, 1965, p. 164, 166.

4. In contrast, we find in the *Lü Shih Ch'un Ch'iu* (a Chinese philosophical compendium of the third century B.C.) the following observation: "If a man climbs a mountain, the oxen below look like sheep and the sheep like hedgehogs. Yet their real shape is very different." See Joseph Needham, *Science and Civilisation in China* (Cambridge at the University Press, 1956), vol. 2, p. 82.

5. Benjamin Lee Whorf, "An American Indian model of the universe," *Collected Papers on Metalinguistics* (Washington, D.C.: Foreign Service Institute, 1952), pp. 47–52.

6. Arthur O. Lovejoy and George Boas, *Primitivism and Related Ideas in Antiquity* (Baltimore: Johns Hopkins University Press, 1935).

7. A. C. Graham, *The Book of Lieh Tzŭ* (London: John Murray, 1960), pp. 34–35.

8. Philip W. Porter and Fred E. Lukermann, "The geography of utopia," in David Lowenthal and Martyn Bowden, eds., *Geographies of the Mind* (New York: Oxford University Press, 1976), pp. 197–223.

9. Rudolf Arnheim, *Art and Visual Perception* (Berkeley and Los Angeles: University of California Press, 1965), p. 240.
10. E. Minkowski, *Lived Time: Phenomenological and Psychological Studies* (Evanston: Northwestern University Press, 1970), pp. 81–90.
11. John T. Ogden, "From spatial to aesthetic distance in the eighteenth century," *Journal of the History of Ideas*, vol. 35, no. 1, 1974, pp. 63–78.
12. Frederick Bradnum, *The Long Walks: Journeys to the Sources of the Nile* (London: Victor Gollancz, 1969), pp. 21–22.
13. Christopher Salvesen, *The Landscape of Memory: A Study of Wordsworth's Poetry* (London: Edward Arnold, 1965), pp. 156–157.
14. Brian Elliott, *The Landscape of Australian Poetry* (Melbourne: F. W. Cheshire, 1967), p. 3.
15. David Lowenthal, "The American way of history," *Columbia University Forum*, vol. 9, 1966, pp. 27–32.
16. Patrick Hart, *Thomas Merton/Monk: A Monastic Tribute* (New York: Sheed and Ward, 1974), pp. 73–74.
17. E. W. Straus, *The Primary World of Senses* (New York: The Free Press, 1963), p. 33.
18. D. N. Parkes and N. Thrift, "Timing space and spacing time," *Environment and Planning A*, vol. 7, 1975, pp. 651–670.
19. Leslie A. White, "The world of the Keresan Pueblo Indians," in Stanley Diamond, ed., *Primitive Views of the World* (New York: Columbia University Press, 1964), pp. 83–94.
20. T. G. R. Strehlow, *Aranda Tradition* (Melbourne: Melbourne University Press, 1947), pp. 30–31.
21. R. M. Berndt and C. H. Berndt, *Man, Land and Myth in North Australia: The Gunwinggu People* (East Lansing: Michigan State University Press, 1970), pp. 18–41.
22. N. I. Wu, *Chinese and Indian Architecture* (New York: Braziller, 1963), pp. 29–45.

[10]

Intimate Experiences of Place

1. Gaston Bachelard, *The Poetics of Space* (Boston: Beacon Press, 1969), pp. 40–41.
2. S. L. Washburn and Irven De Vore, "Social behavior of baboons and early man," in S. L. Washburn, ed., *Social Life of Early Man* (Chicago: Aldine, 1961), p. 101.
3. See, for example, Mario Praz's deep attachment to things. "Things remain impressed in my memory more than people. Things which have no soul, or rather, which have the soul with which we endow them, and which can also disappoint us when one day the scales fall from our eyes; but people disappoint us too, often, for it is only very rarely that we come to know them, and when we think we know them and feel ourselves in unison with them, it is because it is the thickest scales of all which then cover our eyes—the scales of love." *The House of Life* (New York: Oxford University Press, 1964). Quoted by Edmund Wilson, *The Bit between My Teeth* (London: W. H. Allen, 1965), p. 663.
4. Tennessee Williams, *The Night of the Iguana* (New York: New Directions, 1962), Act 3.

5. Augustine, *Confessions*, book 4, 4:9.
6. Robert S. Weiss, *Loneliness: The Experience of Emotional and Social Isola-tion* (Cambridge: The MIT Press, 1973), pp. 117–119.
7. Christopher Isherwood, *A Single Man* (New York: Simon & Schuster, 1964), p. 76.
8. Paul Horgan, *Whitewater* (New York: Paperback Library edition, 1971), p. 163.
9. John Updike, "Packed dirt, churchgoing, a dying cat, a traded car," *New Yorker*, 16 December 1961, p. 59.
10. Robert Coles, *Migrants, Sharecroppers, Mountaineers* (Boston: Little, Brown and Co., 1971), p. 204.
11. *Ibid.*, p. 207.
12. Doris Lessing, *The Golden Notebook* (New York: Bantam Book edition, 1973), p. 390.
13. Wright Morris, *The Home Place* (New York: Charles Scribner's Sons, 1948), pp. 138–139.
14. Freya Stark, *Perseus in the Wind* (London: John Murray, 1948), p. 55.
15. Helen Santmyer, *Ohio Town* (Columbus: Ohio State University Press, 1962), p. 50.
16. Archie Lieberman, *Farm Boy* (New York: Harry N. Abrams, 1974), pp. 130–131.
17. Robert M. Pirsig, *Zen and the Art of Motorcycle Maintenance* (New York: William Morrow, 1974), p. 341.
18. On the meaning and symbolism of the American town, see Page Smith, *As a City upon a Hill: The Town in American History* (Cambridge: MIT Press, 1973); on the courthouse square and small town, see J. B. Jackson, "The almost perfect town," in *Landscapes* (University of Massachusetts Press, 1970), pp. 116–131.
19. On intimate places of general symbolic import see Gaston Bachelard, *The Poetics of Space* (Boston: Beacon Press, 1969), and Otto Bollnow, *Mensch und Raum* (Stuttgart: Kohlhammer, 1971). On the symbolism of the house see Clare C. Cooper, "The house as symbol of self," *Institute of Urban and Regional Development*, University of California, Berkeley, Reprint no. 122, 1974); J. Douglas Porteous, "Home: the territorial core," *Geographical Review*, vol. 66, no. 4, 1976, pp. 383–390.

[11]
Attachment to Homeland

1. John S. Dunne, *The City of Gods: A Study of Myth and Mortality* (London: Sheldon Press, 1974), p. 29; J. B. Pritchard, *Ancient Near Eastern Texts* (Princeton: Prince-University Press, 1955), pp. 455ff.
2. *Ibid.*, p. 85; see also René Guénon, "La Cité divine," in *Symboles fon-damentaux de la science sacrée*, (Paris: Gallimard, 1962), pp. 449–453; Lewis R. Farnel, *Greece and Babylon* (Edinburgh: T. Clark, 1911), pp. 117–120.
3. *Appian's Roman History*, book 8, chapter 12:28, trans. Horace White (Lon-don: William Heinemann, 1912), vol. 1, p. 545.
4. M. D. Fustel de Coulanges, *The Ancient City* (Garden City, New York:

Doubleday Anchor Books, n.d.), pp. 36–37. (First published as *La Cité antique* in 1864.)
5. Martin P. Nilsson, *Greek Popular Religion* (New York: Columbia University Press, 1940), p. 75.
6. De Coulanges, *The Ancient City*, p. 68.
7. Euripides, *Hippolytus*, 1047–1050. See Ernest L. Hettich, *A Study in Ancient Nationalism* (Williamsport, Pa.: The Bayard Press, 1933).
8. Pericles' funeral oration in Thucydides, *The History of the Peloponnesian War*, book 2: 36, trans. Richard Crawley (Chicago: The University of Chicago Press), Great Books, vol. 6, 1952, p. 396.
9. Isocrates, *Panegyricus*, 23–26, trans. George Norlin (Cambridge: Harvard University Press, 1928), vol. 1, p. 133.
10. Raymond Firth, *Economics of the New Zealand Maori* (Wellington, New Zealand: Government Printers, 1959), p. 368.
11. *Ibid.*, p. 370.
12. *Ibid.*, p. 373.
13. Clarence R. Bagley, "Chief Seattle and Angeline," *The Washington Historical Quarterly*, vol. 22, no. 4, pp. 253–255. The speech is reported by Dr. Henry A. Smith in *Seattle Sunday Star*, October 29, 1877. Although the sentiment is Chief Seattle's, the words in English are those of Dr. Smith, whose feeling for rhetoric may have been influenced by classical learning.
14. Leonard W. Doob, *Patriotism and Nationalism: Their Psychological Foundations* (New Haven: Yale University Press, 1952), p. 196.
15. Ernest Wallace and E. Adamson Hoebel, *The Comanches: Land of the South Plains* (Norman: University of Oklahoma Press, 1952), p. 196.
16. Chief Standing Bear, *Land of the Spotted Eagle* (Boston: Houghton Mifflin, 1933), p. 43, pp. 192–193.
17. W. E. H. Stanner, "Aboriginal territorial organization: estate, range, domain and regime," *Oceania*, vol. 36, no. 1, 1965, pp. 1–26.
18. T. G. H. Strehlow, *Aranda Traditions* (Melbourne: Melbourne University Press, 1947), p. 51.
19. *Ibid.*, pp. 30–31; see also Amos Rapoport, "Australian aborigines and the definition of place," in W. J. Mitchell, ed., *Environmental Design and Research Association*, Proceedings of the 3rd Conference at Los Angeles, 1972, pp. 3-3-1 to 3-3-14.
20. Robert Davis, *Some Men of the Merchant Marine*, unpublished Master's thesis, Faculty of Political Science, Columbia University Press, 1907; quoted in Margaret M. Wood, *Paths of Loneliness* (New York: Columbia University Press, 1953), p. 156.
21. Raymond Firth, *We, the Tikopia* (London: George Allen & Unwin, 1957), pp. 27–28.
22. John Nance, *The Gentle Tasaday* (New York: Harcourt, Brace, Jovanovich, 1975), pp. 21–22.
23. *Ibid.*, p. 57.
24. *Tao Te Ching*, chapter 80; quoted in Fung Yu-lan, *A Short History of Chinese Philosophy* (New York: Macmillan, 1948), p. 20.
25. Archie Lieberman, *Farm Boy* (New York: Harry N. Abrams, 1974), p. 36.
26. *Ibid.*, p. 130.
27. *Ibid.*, p. 293.

Notes

[12]
Visibility: the Creation of Place

1. Wallace Stevens, *Collected Poems* (New York: Knopf, 1965), p. 76.
2. Langer, *Feeling and Form*, p. 40.
3. R. M. Newcomb, "Monuments three millennia old—the persistence of place," *Landscape*, vol. 17, 1967, pp. 24–26; René Dubos, "Persistence of place," in *A God Within* (New York: Charles Scribner's, 1972), pp. 111–134.
4. Langer, *Feeling and Form*, p. 96.
5. *Ibid.*, p. 98.
6. John Y. Keur and Dorothy L. Keur, *The Deeply Rooted: A Study of a Drents Community in the Netherlands*, Monographs of the American Ethnological Society, vol. 25, 1955.
7. Maurice Halbwachs, *The Psychology of Social Class* (Glencoe, Illinois: The Free Press, 1958), p. 35. As to village solidarity vis-à-vis outsiders, Paul Stirling notes: "The virtues of the village are an eternal topic of conversation with outsiders, and of banter between men of different villages. Every village has the best drinking water, and the best climate." In "A Turkish village," in *Peasants and Peasant Societies*, ed. Teodor Shanin (Harmondsworth: Penguin, 1971), p. 40.
8. G. William Skinner, "Marketing and social structure in rural China," *The Journal of Asian Studies*, vol. 24, no. 1, 1964, p. 32.
9. *Ibid.*, p. 35.
10. *Ibid.*, p. 38.
11. Herbert J. Gans, *The Urban Villagers* (New York: The Free Press, 1962), p. 105.
12. *Ibid.*, p. 107.
13. Walter Firey, *Land Use in Central Boston* (Cambridge: Harvard University Press, 1947), pp. 45–48, pp. 87–88, p. 96.
14. Caroline F. Ware, *Greenwich Village 1920–1930* (Boston: Houghton Mifflin Co., 1935), pp. 88–89.
15. Eliade, *The Sacred and the Profane*, p. 49.
16. Charles Pendrill, *London Life in the 14th Century* (London: Allen & Unwin, 1925), pp. 47–48.
17. Anselm Strauss, *Images of the American City* (New York: Free Press, 1961).
18. Jan Morris, "Views from Lookout Mountain," *Encounter*, June 1975, p. 43.
19. A. Andrewes, "The growth of the city-state," in Hugh Lloyd-Jones, ed., *The Greeks* (Cleveland and New York: World, 1962), p. 19.
20. Thucydides, ii, 37 (B. Jowett translation).
21. William R. Halliday, *The Growth of the City State* (Chicago: Argonaut, 1967), p. 94.
22. Ernst Moritz Arndt, "Was ist des Deutschen Vaterland?" in *The Poetry of Germany*, trans. Alfred Baskerville (Baden-Baden and Hamburg, 1876), pp. 150–152; quoted in Louis L. Snyder, *The Dynamics of Nationalism* (Princeton, N.J.: D. Van Nostrand, 1964), p. 145.
23. Albert Mathiez, *Les Origines des cultes révolutionnaires (1789–1792)* (Paris: Georges Bellars, 1904), p. 31; quoted in C. J. H. Hayes, *Essays on Nationalism* (New York: The Macmillan Co., 1928), p. 103.
24. Leonard W. Doob, *Patriotism and Nationalism: Their Psychological Foundations* (New Haven: Yale University Press, 1964), p. 163.

25. Hayes, *Essays on Nationalism*, p. 108–109.
26. *Ibid.*, p. 65.

[13]
Time and Place

1. John G. Gunnell, *Political Philosophy and Time* (Middletown, Conn.: Wesleyan University Press, 1968), pp. 55–56, 65–66.
2. "Even the most highly developed nomads do not go far, no more than 150 or possibly 200 miles of farthest distance in the year, and relatively long spells of pitched tents are desired. The women wish it so, caring nothing for floristic composition of the grazing." F. Fraser Darling, "The unity of ecology," *The Advancement of Science*, November 1963, p. 302.
3. Michael Young and Peter Willmott, *The Symmetrical Family* (New York: Pantheon Books, 1973), pp. 148–174, pp. 239–262.
4. James K. Feibleman, *Philosophers Lead Sheltered Lives* (London: Allen & Unwin, 1952), p. 55.
5. Helen Gardner, "Clive Staples Lewis," *Proceedings of the British Academy*, vol. 51, 1965, p. 421.
6. V. S. Naipaul, *The Mimic Men* (London: Andre Deutsch, 1967), pp. 204–205. 205.
7. Eugene Ionesco, *Fragments of a Journal* (London: Faber & Faber, 1968), p. 11.
8. Simone de Beauvoir, *The Coming of Age* (New York: Putnam, 1972), p. 365.
9. Aristides, "The opinionated librarian," *The American Scholar*, Winter, 1975/76, p. 712.
10. "In our society, stuffed animals, particularly the teddy bear, occupy an important role in the life of every toddler. By the time we go to school, however, usually these animals are relegated to the clutter chest in the attic. It is, therefore, of psychological interest to note that college stores sell stuffed animals. Even though the conscious rationalization for the purchase of animals may be because they are 'cute' or bear the college emblem, they still are, of course, nothing more than the old teddy bear who now goes to college, serving as reassurance that nothing has changed." In Daniel A. Sugarman and Lucy Freeman, *The Search for Serenity: Understanding and Overcoming Anxiety* (New York: The Macmillan Company, 1970), p. 81.
11. Claude Lévi-Strauss, *The Savage Mind* (London: Weidenfeld & Nicolson, 1966), p. 234.
12. E. E. Evans-Pritchard, *The Nuer* (Oxford: Clarendon Press, 1940), p. 108. On the Nuer's tendency to treat events as unique and to deny them of historicity see David F. Pocock, "The anthropology of time-reckoning," in John Middleton, *Myth and Cosmos: Readings in Mythology and Symbolism* (Garden City, N.Y.: The Natural History Press, 1967), p. 310.
13. Jiro Harada, *A Glimpse of Japanese Ideals: Lectures on Japanese Art and Culture* (Tokyo: Kokusai Bunka Shinkokai, 1937), p. 7.
14. T. T. Chen, "The Chauchow stone bridge," *People's China*, vol. 15, August 1955, pp. 30–32; Andrew Boyd, *Chinese Architecture and Town Planning* (Chicago: University of Chicago Press, 1962), p. 155.

15. Homer Dubs, *The History of the Former Han Dynasty* (Baltimore: Waverly Press, 1938), vol. 1, pp. 181 and 183.
16. L. S. Yang, *Les Aspects économiques des travaux publics dans la Chine impériale* (Paris: Collège de France, 1964), p. 18.
17. *The Travels of Marco Polo*, trans. R. Latham (Harmondsworth: Penguin, 1958), pp. 98–100.
18. C. S. Lewis, *The Discarded Image* (Cambridge at the University Press, 1964), pp. 182–183.
19. Sigfried Giedion, *Architecture and the Phenomena of Transition*, p. 231.
20. Georges Poulet, *Studies in Human Time* (Baltimore: The Johns Hopkins University Press, 1956), pp. 23–24.
21. Howard Mumford Jones, *O Strange New World* (New York: Viking Press, 1964), p. 359.
22. Alvin Schwartz, *Museum: The Story of America's Treasure Houses* (New York: E. P. Dutton, 1967), pp. 126–127.
23. David Lowenthal, "The past in the American landscape," in David Lowenthal and Martyn J. Bowden, eds., *Geographies of the Mind* (New York: Oxford University Press, 1976), p. 106.
24. In the 1960s there were some five thousand museums in the United States. About half were concerned with history. The remainder were divided more or less equally between art and science. See Schwartz, *Museum*, pp. 29 and 124; Dillon Ripley, *The Sacred Grove: Essays on Museums* (New York: Simon and Schuster, 1969), p. 89.
25. Perhaps the preservationist's most convincing argument rests not on aesthetics and sentiment but on practical result—for instance, the idea that the right kind of preservation can save our disappearing downtowns. See Peirce F. Lewis, "To revive urban downtowns, show respect for the spirit of the place," *Smithsonian*, vol. 6, no. 6, 1975, pp. 33–40; and "The future of the past: our clouded vision of historic preservation," *Pioneer America*, vol. 7, no. 2, 1975, pp. 1–20.
26. "One who has seen the handsome pedestrian mall in the heart of Rotterdam, which the Nazis had bombed out in their wanton attack on Holland, might think that only a thorough bombing would make possible the restoration of the heart of the American city. Students of the city have remarked that one reason for the attractiveness of San Francisco is that it had had the advantage of a devastating earthquake." Herbert J. Muller, *The Children of Frankenstein* (Bloomington: Indiana University Press, 1970), p. 270.
27. Sentiment for the past can be quantified and merchandized. "In affluent countries the merchandising of selective aspects of nostalgia for the cultural past is both possible and profitable. It is suggested here that elements of place and activity nostalgia invest many historical landscape features, and that the exploitation of them is a recreational activity worth definition and measurement." Robert M. Newcomb, "The nostalgia index of historical landscapes in Denmark," in W. P. Adams and F. M. Helleiner, eds., *International Geography 1972* (Toronto: University of Toronto Press, 1972), vol. 1, sec. 5, pp. 441–443.

INDEX

Index

229

Index